The Logic of the Plausible and Some of its Applications

To my friend, François Le Lionnais,
the last of the encyclopaedic minds.

The Logic of the Plausible and Some of its Applications

by
René Leclercq

Plenum Press · London and New York

Library of Congress Catalog Card Number: 74–1618
ISBN 0 306 30793 6

Plenum Publishing Company Ltd
4a Lower John Street
London W1R 3PD
Telephone 01-437 1408

U.S. Edition published by
Plenum Publishing Corporation
227 West 17th Street
New York, New York 10011

MADE AND PRINTED IN GREAT BRITAIN BY
THE GARDEN CITY PRESS LIMITED
LETCHWORTH, HERTFORDSHIRE
SG6 1JS

PREFACE

So simple and imperfect as it may appear this book has made use of knowledge on invention and discovery accumulated during a lifetime. Those persons who would be tempted to emphasize only its imperfections should read the correspondence exchanged between Cantor and Dedekind at the end of the nineteenth century; they would then realize how difficult it was, even for an outstanding man, the creator of the set theory, to propose impeccable results in a completely new field. The field I have chosen here is plausibility.

I have proposed an intuitive, some would say a naïve, presentation as I want to reach as large an audience as possible and because I personally believe that it is easier to axiomatize a mathematical theory precisely than to discover it and enunciate its key theorems. Professor Polya said: "The truly creative mathematician is a good guesser first and a good prover afterward."

For centuries a formalized generalized logic was found necessary and many attempts have been made to build it.

Mine is based on plausibility which covers with precision a wider field than probability and makes the formalization of analogy and generalization possible. As Laplace said: "Even in the mathematical sciences, our principal instruments to discover the truth are induction and analogy."* The examples of application I have chosen are not described in detail. A large field remains for those who should become interested and I have no wish to overburden them with hasty solutions. Each solution implies years of hard work and reflection.
I have thought it interesting to summarize the history of systematic thinking, as it is not well known, and it was necessary to place my endeavour in its natural perspective. My book is small. I have, I know, a natural tendency to concentrate my ideas in a few sentences. To this has been added a real danger that my life could end swiftly. I have had to make haste.
It is impossible for me to thank all those who have helped me in formulating and correcting my ideas. They are too numerous, come from too many countries, and have not realized how valuable was their criticism. Special thanks are due to those who believed in my solitary and continuous effort. I thank my secretary, Miss R. S. Vokes, for her help in making my English understandable, Dr. R. E. W. Maddison, F.S.A., for his corrections of language and some historical data, Dr. Trevor Williams for his many suggestions and constant interest. They are not responsible for my mistakes. In England and elsewhere I found as many friends as in France, where François Le Lionnais and Pierre Piganiol were my constant support. I thank Denise for her constant and loving attention during a specially difficult time.

* Essai philosophique sur les probabilités. *Oeuvres complètes*, vol. 7, p. v.

CONTENTS

Preface		v
Chapter 1	The Pioneers of Systematic Thinking	1
Chapter 2	The Logic of the Plausible	35
Chapter 3	Introduction of Some Important Applications	53
References		73
Index		79

Chapter One

THE PIONEERS OF SYSTEMATIC THINKING

1.1 INTRODUCTION

Logic and heuristics (considered as a science) are a result of systematic thinking as opposed to wandering thoughts. The founders of the movement are so numerous that writing their complete history would be impossible and rather boring. As one distinguishes the peaks in a mountain range I have considered the main pioneers among the founders of systematic thinking. The dangers are to draw the wrong conclusion, to believe too much in the individual contributions, to make an imperfect choice and to forget important contributors. This is a risk I had to take in view of the many advantages achieved: easier reading because it is less abstract and more human; a more schematic view which is easier to remember; special tributes paid to pioneers who have been forgotten. To challenge the usual choice may well lead to criticism. I have not mentioned Copernicus, Kepler, Galileo, Whewell, Claude Bernard, Peano, Hilbert, Poincaré, Marx and many others who were first-class thinkers but not as important as pioneers of systematic thinking as the ones I have chosen. The criteria of choice I have followed are to mention either the persons who were the first to formulate an original idea (unexpected at that time) or those persons who for the first time correctly used a method which was more confused in its beginning. Before describing the lives of the known pioneers I should like to pay tribute to the many forgotten men who made the first main inventions of the world.

1.2 ANONYMOUS INVENTORS

Writing, the idea of natural law, numbers, measurement, solving problems were essential in the building of a science however primitive it might have been. The names of the inventors of these first necessary steps are unfortunately forgotten, probably for ever. Even the countries were these discoveries were made are uncertain. We do not know for sure if writing originated among Summerians around 3000 B.C. or a thousand years earlier in Rumania [1]. We are not sure that the Babylonians were the first to solve algebraic problems around 2000 B.C. [2]. We do not know how far the people of Stonehenge went into the understanding of the movement of the stars [3]. It would, however, be unfair not to pay tribute to the unknown pioneers.

1.3 PARMENIDES OF ELEA AND THE PRINCIPLE OF IDENTITY

Primitive tribes accept the possibility of contradiction [4]. Something may exist and not exist, both at the same time.

The possibility of changing a being into its opposite is even accepted. Later, Heraclitus (acme 504–501 B.C.) considered the world as an equilibrium between opposites [5]. The role of Parmenides was to modify this attitude and to adopt a more positive one. He reacted against the dualistic conceptions.

According to Plato (Parmenides 127A) Parmenides visited Athens with Zeno and met the young Socrates. If we believe Antiphon's account, Parmenides was then about sixty-five, grey haired and good looking. They stayed at Pythodorus' house outside the city walls of Ceramicus. This event places the birth of Parmenides around 515 to 510 B.C. According to Diogenes Laertius (IX, 21–3) Parmenides was a son of Pyres and a pupil of Xenophanes. He came of a distinguished and rich family. He did not follow his master. It was through Ameinias that he became a convert to the contemplative life. He is said by Speusippus to have legislated for the citizens of Elea. He wrote hexameter verses on two very prosaic subjects: "The way of truth" and "The way of seeming". His writing is unfortunately obscure and one has to guess the meaning of his sentences taking into account the ideas of the time. I have accepted Paul Tannery's interpretation [5] modified by G. S. Kirk and J. E. Raven [6].

If one premise is accepted then logic compels the rejection of the other. A thing either exists or does not exist. Ignorant mortals are two-faced because they combine opposites. The way of seeming is the combination of the true way and the utterly false way. Obviously Parmenides is for one truth and against contradiction which leads to "seeming". This progress was necessary. Dialectic is a method of reaching truth, and should not be confused with the result.

1.4 SOCRATES AND DIALECTIC

The existence of Socrates has been questioned [7], but the best way of explaining the testimonies of Xenophon, Plato and the caricatures of Aristophanes in *The Clouds* is to accept his existence [8]. It is more difficult to reject the legend created by Plato. Socrates was an Athenian citizen of moderate means who taught philosophy to the young. He was condemned to death and executed in 399 B.C. at the age of seventy. Xenophon considered Socrates as an ideal man. Plato used him as a mouthpiece for his own opinions, except in the *Apology* which shows him as an historical figure [9]. Socrates appears as a high-minded man, sure of himself, persuaded that clear thinking is at the basis of right living.

For Socrates (and Plato) dialectic was the method of seeking knowledge. The method was not invented by Socrates but he perfected it and used it the right way. To a thesis he opposed an antithesis; comparison between the thesis and the antithesis led to a new thesis to which a new antithesis was opposed . . . He acted as a male-midwife for brains. Centuries later dialectic was revised and perfected by Hegel, Karl Marx, Lenin, Mao-Tse-Tung and many others. It is now used by a great number of persons.

1.5 MENAECHMUS AND PURE AXIOMATIC METHOD [10]

The Academy, founded in 385 B.C. by Plato when he was forty-five, became so famous in less than twenty years that students attended from all parts of Greece. Before 350 B.C. Menaechmus of Proconnesus and his brother, Dinostratus, joined the Academy coming to Athens with

* *

their master, the mathematician Eudoxus of Cnidus, who had been directing a school at Cyzicus (on the east coast of the sea of Marmara). Both Plato and Eudoxus played an important role in the birth of mathematics, but it was Menaechmus who expressed for the first time the now accepted principles of the axiomatic method. He wrote between the years 360 and 340 B.C., but little is known about this very important man.

"Elements" were not unusual at the time. Hippocrates of Chios was the first, according to Eudemus, to have written such a book. Leon, whose origins are unknown, compiled some Elements around 365 B.C. Theudius of Magnesia, who also came with Eudoxus of Cnidus, wrote other Elements at the time of Plato's death (348 B.C.). They consisted of a series of enunciations written in such an order that the first ones served to prove the following ones. Plato's nephew, Speusippus, who followed him as head of the Academy, continued the belief that the primary notions or axioms must express fundamental truth. Conversely, Menaechmus held that they should be just postulates or hypotheses (Proclus, *comm. in Eucl.* lib. I), mathematics being a mere exercise of logic. This is the modern attitude which proved itself very successful in making the study of non-Euclidian geometries legitimate. The best known Elements are those of Euclid which were compiled around 300 B.C. Euclid belonged to the new school of Alexandria.

1.6 ARISTOTLE AND THE INVENTION OF SYLLOGISM

The life of Aristotle is well known [11]. He wrote to King Philip of Macedon that he had spent twenty years with

Plato. He left the Academy at the time of Plato's death ($\pm$ 348 B.C.); and as he was seventeen when he entered it we conclude that Aristotle was born around 384 B.C. There exist an unusually great number of old manuscripts on Aristotle's writings [12].

Aristotle was born at Stagira (Thrace). His father, Nicomachus, was family physician to the king of Macedonia. Around seventeen he came to Athens and became a pupil of Plato with whom he remained until the death of the latter. The nomination of Speusippus to the headship of the Academy, and the destruction of his ancestral home, led Aristotle to leave for Asia Minor with Xenocrates, settling in Assos. They were protected by Hermias, Prince of Atarneus. Theophrastus came from Eresos in Lesbos to study with them. Aristotle married Pythias, niece and adopted daughter of Hermias. After three years in Arsos, Aristotle and Theophrastus went to Metylene in Lesbos where they stayed until 343 B.C. Then Aristotle accepted King Philip's invitation to become tutor of Alexander. Soon after, he heard about the terrible fate of Hermias who had been crucified by the Persians and who had asked as a last favour to tell his friends and companions that he had done nothing unworthy of a philospher. Aristotle composed a dedicatory epigram to Hermias which exists to this day on a cenotaph at Delphi. He was therefore opposed to Demosthenes in Athens who was blackening the character of the deceased. The influence of Aristotle on Alexander, the fate of Greece and the Middle East are difficult to assess, but must have been important.

Around 335 B.C. Aristotle returned to Athens and founded the Peripatetic school. He had been away for thirteen years and was then fifty years old. Speusippus had died in 339 B.C. and Xenocrates had succeeded him. Aristotle was able to proclaim himself as the real successor of Plato.

The Academy was superseded by the new school which was first started in the corridors of the palaestra in the Lyceum and later probably in rooms opposite the gates of Diochares. Aristotle was protected by his friend Antipater, a Macedonian, regent and commander-in-chief in Macedon and Greece. Tradition tells us that Aristotle gave his philosophical lectures in the morning and spoke in the afternoon to a larger public on rhetoric and dialectics. At the death of Alexander in 328 B.C., Antipater was in Asia and Aristotle, in view of nationalistic hate and the attacks of Demosthenes, escaped to Chalcis. He remained there until his death a few months later at the age of sixty-three. In his will he carefully provided for his children and students. His wife had been dead for a long time and he was living with Herpyllis who gave him a son. He nevertheless felt very isolated, sensing a terrifying gulf between himself and his surroundings.

Aristotle's greatest influence was in logic where his supremacy survived for centuries. His most important discovery in logic was the syllogism. It may have originated from the use of classification in biology. The *Organon* comprises six volumes [13]. On reading them one becomes convinced of the experimental origin of logic. A syllogism consists of three parts: a major premise, a minor premise and a conclusion. There are different kinds of syllogisms. The most familiar form is the following:

All men are mortal (Major premise)
Socrates is a man (Minor premise)
Socrates is mortal (Conclusion)

The value of syllogism has been overestimated. It is just one form of the many possible ways of reasoning, but that form had to be discovered.

1.7 THEOPHRASTUS AND SCIENTIFIC CLASSIFICATION [14]

The information we possess concerning the life of Theophrastus comes mainly from Diogenes Laertius' *Lives of the Philosophers* compiled four hundred years after the death of Theophrastus who was born in 370 B.C. We already know his connection with Aristotle. He succeeded him as the head of the Peripatetic School. Whether Theophrastus collected the principles of classification from Plato or from Aristotle matters little; Theophrastus was surely the first one to apply those principles in the way they should be applied in science. He was a voluminous writer. Diogenes gives a list of 227 treatises all written by him, but most of them are lost. According to the same Diogenes, Theophrastus died at the age of eighty-five.

1.8 ARCHIMEDES AND THEORY BASED ON EXPERIMENTS

Archimedes was born around 287 B.C. He was the son of the astronomer Phidias and a relation of King Hieron of Syracuse. He went to Egypt and visited Alexandria where he became a friend of well-known geometers. He organized the defence of Syracuse described by Sylius Athicus and Plutarch, but he was more interested in pure theory than in machines. He was killed by a Roman soldier in 212 B.C. The manuscripts of Archimedes which were lost or forgotten came to light during the Middle Ages [15] and even quite recently (end of nineteenth century) [16]. Not much is known about his life. Besides important works in geometry and the near-invention of integral calculus, Archimedes was the first to use precise observation and experimentation as a basis for mathematical theories [17] his two works *On the Equilibrium of Planes* and *On Floating Bodies* aptly illustrating his achievement. Euclid had done the same in optics before Archimedes, but his basic observations were not good enough. Proposition I in Book II demonstrates the celebrated "principle of Archimedes" relating to the apparent loss of weight by an object immersed in a fluid. There is no doubt about the experimental origin of this principle even if the legend about Eureka and the bath is apocryphal.

1.9 ALHAZEN AND THE EXPERIMENTAL METHOD

It is difficult to pick out the first man who used the experimental method. It must be firmly stated that the experimental method is *not* just the use of experiments to illustrate a theory, or the formulation of a theory explaining a series of observations, it is a *succession* of theories and experiments deduced from the theories. An hypothesis which leads to practical experiments is formulated, experiments resulting from the hypothesis are made, the theory is modified according to the results, new experiments are made for verification and so on. The experimental method is therefore the active interrogation of nature by a combination of operational hypotheses and corresponding experiments. I had for many years believed that Peter of Maricourt (in the thirteenth century) was the first man to have used the experimental method in his study of magnets. It is now obvious to me that Alhazen preceded him by two centuries. The Alchemists are out of the race because their theories are a strange mixture of mysticism (not operational by definition) and strange interpretation. The experimental method supposes a positive attitude.

Very little is known of the life of Abu Ali Mohammed Ibn Al Hasan Ibn Al Haythem, usually known as Alhazen. He was born in Basra, probably in A.D. 965. He spent most of his life in Egypt and died in Cairo in 1039. In spite of his great discovery he is less well known than his contemporary, Avicenna (980–1037). Alhazen's works on optics were translated from the original Arabic into Latin by Frederick Risner of Basle in 1572 under the title *Opticae Thesaurus libri septum, per episcopios.* We also know that he was a follower of Lucretius, the author of *De natura rerum*, and was therefore in favour of a positive attitude. Let us summarize the experiments described by Alhazen. Using a kind of camera obscura he observed several candles, when he displaced them their images were displaced, when he suppressed one of them the corresponding image disappeared. He concluded that light follows a straight line and that no mixing occurs in the air. He also observed the reflection of light on all kinds of mirrors: plane, concave, convex, spherical, cylindrical . . . and studied carefully the formation of images. He accepted Euclid's law of reflection but added a proof by analogy by comparing light to small spheres of metal which bounce on a smooth surface making an angle equal and symmetrical to the angle of impact. He explained diffraction by decomposing the velocity of light into normal and parallel components, the normal component being more reduced in less diaphanous bodies. He was centuries ahead of his time and it is not surprising that he was not understood and was completely forgotten for centuries. This is the fate of real pioneers [18].

1.10 RAMON LULL AND MECHANICAL REASONING

Long before Lull, Democritus' sphere was used by astrologers for determining the fate of sick persons [19]. The answer was achieved by the combination of figures in a series of tables. It was a mechanical way of thinking but a static one. In his *Ars Magna* Ramon Lull used, for the first time, rotating concentric circles bearing inscriptions, in such a way that the combination of the inscriptions gave the answers to questions put to the instrument [20]. The life of Ramon Lull is known. He was probably born in 1232, son of one of the Catalans who conquered Majorca in 1228. His family lived in Majorca from 1231. Ramon belonged to the court of James I of Aragon. He married young and had two children. His life was rather dissolute; he was interested in women more than anything else. He radically changed when he realized that a lady he was pursuing with assiduity was suffering from a dreadful form of cancer. He then began to devote his life to religion. This seems to have occurred around 1265. He studied and meditated and from 1271 he travelled, wrote and lectured. He went to Rome, Montpellier, Paris, Pisa and Naples. He tried to convert Muslims in Tunisia where he was killed in 1315; he was then eighty-three years old. Lull was extreme in everything being a strange mixture of mysticism and rational reflection. He wrote very many manuscripts on the subject which interests us:

Ars Universalis written in 1272 in Majorca.
Liber principiorium philosophiae written between 1272 and 1275.
Introductoria Artis demonstrativas, written in Montpellier in 1283.

Ars inveniendi particularia in universalibus, written in Montpellier in 1283.
Ars inventiva veritatis, Montpellier 1287.
Lectura compendiosa tabulae generalis, Naples 1292.
Lectura Artis inventivae et tabulae generalis, Rome 1295.
Ars generalis ad omnes scientias, Montpellier 1304.
Ars demonstrativa, Paris 1309.
. . .

He preached his ideas with passion and was never completely forgotten. In 1685 Leibniz was writing about him: "Raymond Lull also studied mathematics and in a fashion discovered the art of combinations. Lull's art would undoubtedly be a wonderful thing if those fundamental terms of his, Goodness, Magnitude, Duration, Power, Wisdom, Will, Virtue, Glory, were not so vague, and consequently serve only to express but in no way to discover the truth." As we shall see, Leibniz believed in the possibility of an *Ars inveniendi* based on combinations, as in fact do several modern writers.

1.11 MAUROLICO AND MATHEMATICAL RECURRENCE

One of the most important forms of reasoning in mathematics is recurrence, which is considered by philosophers as a perfect form of induction. If a property is strictly related to the value of a variable; if the property is verified for at least one value and if one can prove that the property is true for the variable equal to n (n = any value) it is then true for the variable equal to the successor of n; then the property is general. The basis of recurrence is the formation of numbers, each number having one successor, and one only. Recurrence was used, most probably for the first time, by Francesco Maurolico. He was born in Messina (Sicily) in 1494 of Greek parents. He became a priest and professor of mathematics. He is sometimes mentioned for his many translations into Latin of classical Greek authors: Euclid, Appolonius, Archimedes . . . and for a remarkable book on physics. His *Opuscula mathematica* in which recurrence is used is now a very rare, and completely forgotten, book. He died in 1575. He is very seldom mentioned even in specialized works.

1.12 FRANCIS BACON AND THE ORGANIZATION AND POLICY OF RESEARCH

St. Albans is a small and quiet city, successor to the Romano-British settlement of Verulamium, easily reached from London. In St. Michael's church there is a statue erected to the memory of Francis Bacon. Very few people are interested in this. The Roman museum, across the way, seems to be a much greater attraction.

Francis Bacon was born on January 22, 1561 at York House in the Strand, London. He was educated in the great household of his uncle, Lord Burghley, and spent three years at Trinity College, Cambridge. Son of Nicholas Bacon, Lord Keeper of the Seal, he was sent to be attached to the English ambassador to France. He was then already very much engrossed in science and became interested in the criticism of Aristotle by Peter Ramus. Bacon's own development in inductive logic was influenced by Ramus. Back in England, Francis was admitted to Gray's Inn, and later studied law to practise it as a profession. All his life he pursued professional, philosophical and political activities simultaneously. He entered the House of Commons in 1584. By courting Queen Elizabeth and subsequently King James I and important officials, he finally succeeded in being appointed Lord Chancellor of his country and was created Viscount St. Albans. He had to struggle to get enough money to live according to his rank and was condemned for accepting bribes. He did not marry until he was forty-six, his wife being very much younger; he never seemed to be in love with her. Francis Bacon died on April 9, 1626. He had been a very ambitious and highly rational person [21].

Francis Bacon was in favour of the experimental method which he described in *The Dignity and Advancement of Learning* in 1605. He thought that hypothesis could be formulated logically from available observations by the use of an inductive method. He therefore varied the factors connected with the existence of a phenomenon and observed the effects. Unfortunately Bacon was not a good experimenter and did not convince his contemporaries.

Bacon was an indisputable pioneer in the organization of research and in his belief in the interest of a policy of research. His ideas are expressed in *Nova Atlantis*, a book he wrote in 1620 and which was never finished [22]. It is a sort of science-fiction novel written in Latin. Bacon imagined a perfect research organization in an island where research was used on a grand scale for improving the welfare of the population. The unusual combination in Bacon of a statesman and a scientist produced, three hundred years before its application, a theory which is now accepted the world over.

1.13 RENÉ DESCARTES AND KNOWLEDGE CONSIDERED AS A SEARCH FOR ORDER

René Descartes was born in La Haye in Touraine on March 31, 1596. His father was a councillor of the Parliament of Brittany, who left him a modest amount of landed property. From 1604 to 1612, René Descartes was educated at the Jesuit College of La Fèlche where he learned modern mathematics. He was much impressed by the logical order and the rigour of the proofs of mathematics compared with those of metaphysics. In 1612 he went to Paris where, finding the social life impossible, he quickly retired to a secluded place in the Faubourg St. Germain to study geometry. He enlisted in the Dutch army in compliance with his father's wishes and to learn "from the great book of the world". As Holland was at peace he in fact spent two years in undisturbed meditation. The coming of the Thirty Years' War led him to enlist in the Bavarian army in 1619. During the winter of 1619–20 he had a visionary revelation and discovered his vocation. In 1621 he quitted the military profession and went to Italy, later returning to Paris. Being disturbed again by Parisian life he joined the army which was besieging La Rochelle, and finally settled in Holland. He lived there for twenty years, except for a few visits to France and one to England. He never married but he had by a maid-servant a natural daughter who died at the age of five. This was the greatest sorrow of his life. He always worked for only a few hours a day, without using many books, but was able to concentrate intensely. He wanted to live quietly and to avoid being in trouble with the Church. Even in Holland he was subjected to vexation by protestant bigots. He was fortunately helped by the French ambassador and the Prince of Orange. Invited by Queen

Christina, who sent a warship to fetch him, he went to live at the Swedish court. Unfortunately Christina could not spare the time for her daily lesson except at five o'clock in the morning. Being a delicate person, used to sleeping until nearly midday, Descartes could not withstand the bitter Scandinavian winter and died on February 11, 1650. His body was brought back to France and buried in the Church of St. Geneviève. It was finally transferred to St. Germain-des-Près during the French revolution [23].

The most celebrated book of Descartes is *Le Discours de la méthode** where he uses his method of thinking: rejecting authority as a source of knowledge, accepting evidence of elementary well observed data as a criterion for truth, building from the simple to the complex, arranging the demonstrations in an orderly and progressive fashion. He also imagined a geometrical representation of algebraic equations [24]. The rules for applying order in any domain are best expressed in his book *Rules for the Direction of Thinking* written in Latin and never completed. The manuscript was sent by Chanut, the French ambassador to Sweden and a friend of Descartes, to Clerselius in France. It was published in Amsterdam in 1701.

Descartes was right in believing that knowledge is synonymous with ordering data. Knowledge and life are order as opposed to the apparent disorder of day to day facts.

1.14 PASCAL AND FERMAT, A MATHEMATICAL THEORY OF PROBABILITY

Pierre de Fermat was born near Montauban (France) probably in 1601. He was councillor of the Parliament of

* "Discourse on method."

Toulouse and died in Castres in 1665. He never taught mathematics but was a gifted amateur. His arithmetical discoveries are outstanding [25].
Blaise Pascal was born at Clermont (France) in 1623. He was unusually precocious. He became a first-class mathematician, a very good experimenter and the discoverer of many mathematical relationships. As an experimenter he wrote a treatise on the void which is a model of experimental method. He became mystical and wrote *Les Pensées* which is a model of conciseness and profundity. He died in 1662 [26].
If Cardano (1500–76) was the first to believe in the possibility of a theory of chance, he was unable to realize it. In an exchange of correspondence between Fermat and Pascal in 1654 they both tried to solve a problem about gambling which had been proposed by de Méré. In that correspondence probability is defined as the quotient obtained by dividing the number of favourable cases by the number of possible cases, a definition which was retained as fundamental for many years. It must be added that the possible cases must be equally probable. Probability later became a science. It was prefected by Huyghens (1629–95), Jacques I. Bernouilli (1654–1705), Abraham de Moivre (1667–1754), Bayes (1671–1746), Laplace (1749–1827), Legendre (1752–1833) and Gauss (1777–1855)... [27].

1.15 LEIBNIZ AND THE IDEA OF MATHEMATICAL LOGIC

Gottfried Wilhelm Leibniz was born in Leipzig on July 1, 1646. His father died when he was only six and he inherited a large library which made him mostly an autodidact. His formal education was strictly traditional:

German-Protestant-Aristotelian. He owed, however, a great deal to two of his teachers at Leipzig: Adam Scherzer and Jacob Thomasius. Dissatisfied with the teaching of mathematics at Leipzig he spent the summer of 1663 at Jena studying under Erhard Weigel who was in favour of applying the rigour of the mathematical method to the whole of philosophy. Leibniz developed the same idea later. In 1666, Leibniz wrote *De arte combinatoria* in the hope of obtaining a position as a teacher in Leipzig. It is an immature book but it contains some of the basic ideas of Leibniz' future logic and the idea of "an alphabet of human thought". The divisions are traditional: the logic of discovery and the logic of judgement are successively developed. His endeavour failed and he turned his back on university life, rejecting an offer of professorship in 1667. He obtained a degree of Doctor of Laws and elected to be in direct contact with political power. He eventually became historian to the house of Brunswick. He travelled extensively and wrote on practically every subject. He spent four years in Paris where he met Huyghens and many other mathematicians and physicists. He was in favour of the union of Christian churches. Independently of Newton he invented the integral calculus. He died very lonely and temporarily forgotten in 1716 [28].

Leibniz tried to reduce traditional forms of inferences to an algebraic calculus [29]. All A is B is represented by $A = AB$. If $A = AB$ and $B = BC$, then $A = AC$.* A syllogism in Barbara corresponds to $B = BC$, $A = AB$, then $A = AC$. He carried on using this kind of syllogism but he did not introduce the modern operations. Believing in the possibility of finding the alphabet of human thought,

* An example of syllogism in Barbara, one of the classical forms, is described in §2.3.1.

he tried to reduce the logic of discovery into a combination of key concepts which he was never able to achieve, and it would probably have been a failure as have been similar ones made since. It is indeed impossible to reduce knowledge to a finite number of axioms, as Godel demonstrated much later. The interesting part in Leibniz' endeavour lies in his belief in an algebraic language for logic and in the possibility of applying such a language to all aspects of reasoning.

1.16 LEIBNIZ AND NEWTON AND THE RESTORATION OF CONTINUITY BY INTEGRAL CALCULUS

It was an old dream to be able to integrate infinitely thin sections of a volume into the complete volume, to integrate a geometrical relation valid for any point of a curve to the whole curve ... Archimedes had already found the solution for a simple case. Fermat was interested in the same subject and suggested a promising approach. Pascal was also interested in the problem. It was through reading Wallis and Fermat that Newton turned his attention to the subject and it was through reading Pascal that Leibniz found his own solution. The discovery was made independently by the two men. The charges of plagiarism brought against Leibniz by Newton and his partisans are unfounded [30]. Newton made his discovery progressively. Traces of it can be found in a short paper of 1665–6 and in *De Analysi* (1669). The statement of the binominial theorem appeared only in 1776. The discovery of Leibniz came later but his symbolism and method were more practical and obviously independent. He described his method in a letter to Newton of 1677. At the time relations between the two men were good. Leibniz was always frank and ready to recognize Newton's priority. Newton remained secretive but polite. He later accepted the suggestion of plagiarism made by his friends. Leibniz' publication "Nova methodus pro maximis et minimis", dated 1684, appeared in *Acta Eruditorum*. The method of Newton was published in abbreviated form in 1693 by Wallis in volume II of his *Collected Works*. The book Newton himself wrote on integration was not published (translated from Latin) until 1736.

Isaac Newton was born on December 25, 1642 in Woolsthorp Manor. His father and ancestors were yeoman farmers. His father had died before he was born, and it was fortunate that his mother strongly approved of education. At the age of eighteen he went to Cambridge where he was a sizar or a student who paid his way through college by waiting on his tutor (31). He studied arithmetic, Euclid, trigonometry and the Copernican system. In Cambridge, Newton came into contact with Isaac Barrow, a well-known mathematician. In 1665 he took his degree of Bachelor of Arts without any particular distinction. In the same year the great plague spread from London to the country and the University of Cambridge was closed. He worked alone for nearly two years and laid down the principles of three fundamental discoveries: integral calculus, the theory of light and of universal gravitation. Much later, when he was asked how he made those discoveries, he said: "By always thinking unto them." In 1668 he became a Major Fellow of Trinity College and a year later Lucasian Professor of Mathematics. He was only twenty-six. He experimented with prisms and lenses and constructed a reflecting telescope. Among his main works are the *Principia Mathematica* published in 1686, in which he used the axiomatic form, of analysis. His *Opticks* was published in 1704. The origins of his *Principia* have been traced through a careful study of the documents left by him [32]. Newton went to London as Warden of the Mint in 1696 and became Master of the same in 1669. In 1703 he became President of the Royal Society. Newton was also interested in theology and may be considered as the founder of Unitarianism. He died in 1727. He was a very unusual man, shy and irritable. He preferred to be alone and was reluctant to publish. He was able to concentrate for a great length of time and had a high opinion of himself.

1.17 MATTHEW HALE AND THE PLAUSIBLE REASONING [33]

Logical thinking is used in justice in order to analyse human situations, to discover culprits, to apply the law and to convince juries. A good lawyer should be a good logician and indeed some of them have been among the best. Some have even played an important role in the progress of systematic thinking. Francis Bacon and Leibniz are among those few. Matthew Hale (1609–76), Lord Chief Justice of England under Charles II, was engrossed in scientific research. He was a close friend of John Wilkins, prominent in the founding of the Royal Society and many members of that new society were his intimates. He might have acquired his scientific interest at Magdalen College, Oxford, where Wilkins also studied. Hale studied mathematics and became very conversant with philosophy and "in all curious experiments and rare discoveries" of his age. He studied anatomy and medicine. In 1673, he published *Essay touching the*

Gravitation of fluid bodies and in 1674 *Observations touching the Toricellian experiment*. He did not, unfortunately, make any valuable contribution to experimental science. He merely repeated what was already known and reasoned on it. He was better at theory of science and tried to describe the mental process of invention and discovery. He also published "the analysis of law". He thought that analysis provided a good start, in spite of the fact that the complexity of the law would not permit the reduction of legal thinking to an exact logical method.

The doctrine of relative certainty was used in an early form by several sixteenth-century legally trained French humanists such as Melchior Cano, François Baudouin and Jean Bodin. The great religious controversy of the seventeenth century was useful in clarifying the method. In England, William Chillingworth considered three levels of certainty: the first one available to God only, the second corresponding to complete human certainty, and the third being a reasonable certainty. This primitive approach was perfected by Joseph Glanvil and John Wilkins, but it was Matthew Hale in *The Primitive Origination of Mankind* (1677) who presented the first practical theory. He studied the value of witnesses and the value of evidence. He distinguished four kinds of certainty: the certainty of logical demonstration—the mathematically demonstrated theory—the sensible evidence, and the facts not immediately objected to by our senses. The work of Matthew Hale was followed by the book of Sir Geoffrey Gilbert and by the *Essay concerning human understanding* of John Locke. The first mathematical theory of plausibility was formulated by George Polya [34] who unfortunately used the theory of objective probability in a field which is quite often subjective.

1.18 LEGENDRE AND GAUSS AND SCIENTIFIC MEASUREMENT [35]

Adrien-Marie Legendre was born in Toulouse on September 18, 1752. He was educated at the Collège Mazarin in Paris and became Professor of Mathematics at the Ecole Militaire in Paris from which he resigned in 1780. In 1782 he was awarded a prize by the Berlin Academy for an essay on ballistics. He perfected Euclidian geometry, making it more rigorous; and worked successfully on the theory of numbers and on elliptic functions. His discovery of the theory of least squares was formulated in 1805 in connection with a theory of the orbits of comets. As he was against the royalist government he was deprived of his pension and spent his last days in poverty, dying in Paris on January 10, 1833.

Karl Friedrich Gauss, son of a day labourer, became the prince of mathematics. He was born at Braunschweig on April 30, 1777. He was first educated by his mother and uncle. His abilities showed so early that he was sent to the Carolineum and then to the University of Goettingen. He wrote a theory of numbers, discovered the non-Euclidian geometries and conceived the theory of the least-squares. Legendre and Gauss discovered the same theory completely independently. Gauss was also an astronomer, a geodesist and a physicist. He was an excellent linguist as well. He died in Goettingen on February 23, 1855. His life had been very quiet and very fruitful. He was not very keen on publishing and consequently some of his works lost priority because of his reluctance to fight for new ideas.

The theory of least squares and the Gaussian distribution may be considered as the basis of scientific measurement, and measurement is the basis of good experimentation.

1.19 NICOLAS SADI CARNOT AND THE QUANTIFICATION OF THERMODYNAMICS

Nicolas Léonard Sadi Carnot was born on June 1, 1796. His father was then one of the "Directeurs" of the French Republic and a former officer of the engineers who published in 1786 an "Essai sur les machines" which foresaw the great achievement of his son. The name Sadi was chosen as homage to the great mediaeval Persian poet. The mother of Sadi was very cultivated and a good pianist. Sadi became a pupil at the celebrated "Ecole Polytechnique" in 1812 when he was sixteen. In 1814 he fought heroically in Paris, with the other students of his school, against the allied invaders. In May of the same year he travelled to Antwerp where his father had been the successful defender of the town. A statue was erected by the Belgians in 1865 to Lazare Carnot. When his father was banished by the French royalists in 1815, Sadi, who was a second-lieutenant in the engineers, remained in France at his father's request. In 1819 he was promoted staff-lieutenant in spite of his republican origins and he finally resigned his commission to study. In 1821 he paid a visist to Magdeburg to be with his dying father. In 1824 he published his celebrated book: *Réflexions sur la puissance motrice du feu* [36]. It was the result of several years of meditation which he presented in a simple form. The result was that the scientists of the time did not realize the importance of his work. Son of a revolutionary, he was involved in the French revolution of 1830. He was also a member of the "Association Polytechnique" organized for the popularization of knowledge. Overworked, he became ill in June 1832 and, not yet recovered, was struck down by cholera. He died on August 24, 1832 at the age of thirty-six [37].

The publication of Carnot's book had no immediate repercussions. Contemporaries did not realize its significance. Clapeyron and Poggendorf, from 1830 to 1840, explained it to the scientific world. Before Carnot, steam engines existed, but no good theory had been found to explain the conditions in which heat can be transformed into motion. A new language and new principles had to be formulated. This led to thermodynamics, and much later to the essential theory of information.

1.20 EVARISTE GALOIS, THE GROUP THEORY AND INVARIANCE

Evariste Galois was born on October 25, 1811. His father, who was a headmaster, became mayor of Bourg-la-Reine during the hundred days of Napoleon's return. He was afterwards victimized by Royalists and the local priest and committed suicide. His mother taught him Latin. He studied at Saint-Louis, the well-known French lycée, and read the works of Legendre and Lagrange by himself. He twice failed the "Ecole Polytechnique" examination and the examiner, Dinet, decided that he was not good enough in mathematics! Dinet, otherwise, would never have been remembered! In 1829, Galois was accepted into the "Ecole Normale Supérieure". In 1830 he presented a memoir to the Academy of Science, but the memoir was lost. A little later he proposed another for a prize in mathematics but the prize was awarded to Jacobi and Abel. In July, 1830, Galois opposed the director of his college and he was expelled. He than became a private teacher in mathematics. Being a strong opponent of the Royalist regime he was finally arrested and condemned to prison in 1831. At the beginning of

1832 he was challenged to a duel in which he was severely wounded, and died on May 31, 1832. In the manuscript he wrote before his death he included this desperate sentence: "I have not enough time." The crucial discovery of Galois, the now celebrated group theory, was lost for many years. He was in fact far ahead of his time. Finally, Liouville came across his work and published it with comments. Galois was a very proud and uncompromising young man and became an obvious victim of the establishment [38].

The idea of group theory which was formulated in connection with the solution of linear equations was almost found previously by Lagrange in his treatise on numerical equations. The extension of the idea led to invariance. Invariance became of tremendous importance when Klein, in his Erlangen programme, showed that each geometry could be defined by a specific group of transformation which left the geometrical figures unchanged. Any scientific law is an application of invariance.

1.21 AUGUSTE BRAVAIS AND CORRELATION

The invention of correlation, which is one of the most important relations between phenomena and facts, has been wrongly attributed to Galton, to Pearson and to many others. It was discovered by Auguste Bravais, a French physicist and crystallographer who was born in 1811 and died in 1863. A former student of the "Ecole Polytechnique" he was in many ways ahead of his time, but could not explain his discoveries in very simple language. His *Etudes cristallographiques* of 1851 were understood only in 1879 when explained by Mallard (1833–94). His invention of correlation was published in

Mémoires présentés par divers savants (II, ser. T9, 255) in 1846. Practically nothing is known about the man himself.

Quetelet who published the *Essai de physique sociale* in 1835 did not use correlation. The work of Galton on hereditary genius was published in 1862. Mendel did not employ the correlation coefficient in his celebrated communication of 1865.

1.22 GEORGE BOOLE AND THE OPERATIONS OF LOGIC

George Boole was born in Lincoln, England, in 1815. His father was a shopkeeper, not a very favourable position to ensure his son's education at a time so well described by Charles Dickens. When the same proud father published some Latin verses by his son, George, they were both laughed at by the nobility; but father and son were both very obstinate. George Boole became a teacher, one of those familiar teachers so well sketched by Dickens. After four years of difficult times Boole, who luckily enough for us could not become an officer in the army or a barrister, founded a school. He wanted to teach mathematics as they should be taught, because he did not find anything valuable in the books usually employed at that time. He therefore learned French, German and Italian and read the great mathematicians of the time. It is hard to imagine the kind of effort he had to make. If you want to have a better idea of what it means open *La mécanique céleste* or *La mécanique analytique* of Laplace, which are among the books quoted in connection with Boole. George Boole soon discovered a theory of invariance and Gregory agreed to publish it in the

Cambridge Mathematical Journal. The logician De Morgan, who perfected syllogism, was at that time under strong attack by the Scottish philosopher, William Hamilton (not to be confused with the mathematician William Rowan Hamilton). To help De Morgan, Boole published in 1854 "An investigation of the laws of thought on which are founded the mathematical theories of logic and probabilities". It was the start of modern logic but the book had to be paid for by a friend and himself, and it sold very badly. In 1855, George Boole married. Since 1849 he had been a university professor at Queen's College, Cork, a new college just opened and scarcely known. On December 4, 1864 he unfortunately died of pneumonia contracted as a result of being too devoted to his teaching. The name of George Boole is now known the world over but, as far as I know, his biography has never been published ([39].

Boole wanted to write a complete symbolic logic comprising probability reasoning. He realized only part of it: the introduction of logical operations as a direct extension of mathematical operations. He was probably afraid of writing the now familiar equality: $1+1=1$ which was later proposed by Jevons.

1.23 DMITRI MENDELEEV AND THE USE OF MATRICES OF DISCOVERY

Dmitri Mendeleev was born at Tobolsk on February 8, 1834. He was the youngest in a family of seventeen children. His father, a former professor, was blind and died when Dmitri was only thirteen. Luckily enough his mother was cultivated, intelligent and very energetic. In spite of many difficulties he became a student at the

Pedagogic Institute of St. Petersburg. The discipline of the school was ridiculous but the professors of mathematics and chemistry were good. In June, 1855 Mendeleev left school to become a teacher at secondary school level. The same year he got a master's degree from the University of St. Petersburg. He already had in mind his celebrated classification of the chemical elements. In 1857 Mendeleev became lecturer at the University. Later he spent some time at the University of Heidelberg, in 1865 becoming doctor of science and professor at the University. On February 17, 1869 Mendeleev achieved a first version of his chemical table. He classified the elements and their known properties in a matrix and was able to forecast several unknown elements as well as their main properties. It is truly what is now known as a matrix of discovery. His ideas were not received with enthusiasm. He was challenged and later his discovery was contested. He died on February 2, 1907 having finally found happiness in a second marriage.

1.24 RONALD A. FISHER AND PLANNING OF RESEARCH

Ronald A. Fisher was born in 1890 and died in 1962. He was a Cambridge professor and consultant at the Rothamsted Experimental Station (for agriculture). His books: *Statistical Methods for Research Workers* (1925), *The Design of Experiments* (1935) and *The Statistical Tables* (1938) were very influential. He was very much helped by Frank Yates who was his collaborator for thirty years. "From Fisher came the broad development of many fields of statistics and notably the principles of the design and analysis of experiments. To Yates is mainly due the systematic elaboration of designs to meet a wide range of experimental situations" [40]. Ronald A. Fisher was not the easiest of men; being very engrossed in his own ideas he disagreed with Karl Pearson, and the two great men were opposed to each other. They were two individualists of the first rank [41].

1.25 GÖDEL AND THE LIMITS OF THE AXIOMATIC METHOD

The combination of an improved axiomatic method, the new logical operations and a better symbolism was so successful that one was led to believe in a possible drastic simplification of knowledge which could be reduced to a few essential factors. Einstein searched for the unitary equations, so did Eddington [42]. In 1931 Kurt Gödel, who was then twenty-six, formulated a revolutionary theorem which can be more or less popularized by saying it is impossible to reduce all mathematical knowledge implied in Bertrand Russell's *Principia* into the combination of a limited number of axioms [43]. More generally it was soon proved that knowledge cannot be reduced to the elaboration of a limited number of fundamental roots. One should be careful not to give any mystical value to Gödel's discovery. He did not prove that knowledge was forever mysterious and that systematic invention was impossible. The whole idea is in agreement with the second law of thermodynamics. Entropy defines the level of order in a system, and if entropy increases in the part of the universe we know, that part is in a decreasing order. The world is not in a perfect state and cannot be described in a simple manner. Born in Brno, Czechoslovakia, on April 28, 1906, Ph.D. Vienna 1930, Sc.D.

Harvard 1952, Gödel went to the United States in 1940 and was naturalized in 1948. He is a member of the Institute of Advanced Study, Princeton.

1.26 FRITZ ZWICKY AND THE SYSTEMATIC SOLUTION OF A PROBLEM

Born at Varna in Bulgaria on April 14, 1898 of Swiss parents, Fritz Zwicky first studied at the Polytechnic School of Zürich. He became professor of astrophysics at the Technological Institute of California and director of research for the Aerojet Engineering Corporation at Asuza. He is very much interested in interspatial navigation and was the first to send a projectile away from the earth's gravity. He has also discovered supernovae. An original thinker and very positive as regards the value of his ideas he has made very good friends and a few enemies. I have been a friend of his since 1957 and really enjoy my contacts with him. He calls his method of solutions of a problem, formulated in 1943: "the morphological method" [44]. Zwicky, who has now retired, divides his time between California and Berne. He devotes a great deal of his energy to charitable organizations.
He starts by generalizing the problem to be solved and by enunciating it in such a manner that the solution is not imposed by the vocabulary used. He then analyses the terms of the problem and decomposes it into essential parts, at the same time determining the main parameters. He examines the parts obtained and enunciates the possible solutions for each part. He combines the partial solutions ensuring cohesion between them, choosing the best solution. The danger of the morphological method, which is very general, is the finding of too many solutions,

which Zwicky suggests limiting by restriction of the enunciation. I have proposed the use of axiomatic analysis [45]. The morphological method is simple and constitutes a splendid example of systematic approach to a problem.

1.27 PARTISANS OF NON-SYSTEMATIC METHODS

It would be wrong to believe that everyone considers systematic thinking as the best approach to invention and discovery. By showing the limits of axiomatics Kurt Gödel already belongs to the opposite camp. The French philosopher, Henri Bergson, criticized analysis which, according to him, was incapable of restoring the continuity of reality. More recently, a movement has been growing, especially in the United States, in favour of unsystematic thinking. In 1953, Alex F. Osborn, published *Applied Imagination* [45]. While he is in favour of teaching creativity which supposes some systematization, he is also in favour of using psychological means to stimulate the brain and he proposes disorderly association of ideas of all kinds. The result has been brainstorming. About the same time Gordon stressed the interest of analogical thinking. This was in no way new, but Gordon proposed a team procedure where analogy was the leading process of thinking. Once again this was systematization of a process which tried to get away from the usual systematic thinking. Auto-hypnosis has been used to elicit a response from the sub-conscious. This leads to images which must be interpreted symbolically. Finally, de Bono has proposed lateral thinking as opposed to systematic thinking [46]. Because logic has been identified with deduction and because deduction is not a powerful source of invention, many ways have been suggested which appeal to the unusual, the non-systematic, to improve creativity. This tendency is by no means imperfect as it has led to some practical results, but it would be wrong to confuse creativity with chaos as unfortunately has often been done.

Chapter Two

THE LOGIC OF THE PLAUSIBLE

The logic of the plausible is the transmission of plausibility by symmetry or transitivity, wherever it is correct for it to be done.

2.1 FUNDAMENTAL DEFINITIONS

I shall assume the ideas of *concepts* and *sets* are known. They express objects or collections of objects, also called *points*. They have been the source of lengthy discussions.
A proposition is the expression of *relations* existing between concepts. A concept may vary in correlation with another; it may be contained in another . . .
The *extension* of a set is the number of points of the set.
The *intension* of a set is the enunciation of the propositions in which the set is involved. They express the properties of the points of the set.

The *probability* of an event is the chance that this event has to occur. It is normally expressed by a percentage:

$$\frac{\text{number of favourable cases}}{\text{number of equally possible cases}} \times \frac{100}{1}$$

Objective probabilities are probabilities expressed by a figure on which there is universal agreement. This figure is obtained by a scientific method: mathematical reasoning or statistics. For instance, the probability of obtaining a six by throwing a die is $\frac{1}{6}$. *In the case of objective probabilities, the probability of event* A + *probability that* A *will not occur* (or $p\bar{A}$) = 1. This property is usually summarized by the expression $pA + p\bar{A} = 1$.

Subjective probabilities are personal estimations of probabilities. They are *not* obtained through a rigorous method [47]. One *cannot* write in this case: $pA + p\bar{A} = 1$.

I call *plausibility* the value of a statement determined by a scientific method, or estimated from the value of the method used to obtain the statement. A table of these values can easily be made. A quantitative scientific method will, for instance, give the value between 0·7 and 0·9 in absolute value, according to the number of verifications made. A qualitative subjective method will always give a value under 0·5.

In the case of *measurements* plausibility may be determined two ways. Either by the use of the law of t, discovered by Student; or by the use of the possible error made by using the method chosen. The law of Student gives the probability of a set of observations for which the number of observations and the precision are known [48]. The value of that probability is taken as the value of plausibility. The possible error can be calculated as follows. If m is the mean value of n observations: $m_1, m_2 \ldots m_n$ made by using a method M, one can write:

$$r_1 = m - m_1,\ r_2 = m - m_2 \ldots r_n = m - m_n$$
$$\text{and } m = \frac{m_1 + m_2 + \ldots + m_n}{n}.$$

If M is the real value of the object observed, one can write:

$$R_1 = M - m_1,\ R_2 = M - m_2 \ldots R_n = M - m_n.$$

h' is the precision calculated from $m_1, m_2 \ldots m_n$; h is the real precision. It is easy to prove that m is the most probable value of M. It can be written:

$$E = M - m = \text{Error made by using } m \text{ instead of } M.$$
$$= (M - ri) + (ri - m) = Ri - ri \text{ for } i = 1, 2 \ldots \text{or } n.$$

and also:

$$E = M - \frac{m_1 + m_2 + \ldots + m_n}{n}$$
$$= \frac{n \,.\, M - (m_1 + m_2 + \ldots + m_n)}{n} = \frac{\sum Ri}{n}$$
$$\frac{1}{2p'^2} = \frac{\sum(m - mi)^2}{n} = \frac{\sum ri^2}{n} = \frac{\sum(Ri - E)^2}{n} = s$$
$$= \frac{\sum Ri^2}{n} + E^2 - \frac{2E}{n} - \sum Ri$$
$$= \frac{1}{2p^2} + E^2 - \frac{2E}{n}(nE + \sum ri) = \frac{1}{2p^2} - E^2$$

Using h' instead of h is therefore supposing a precision superior to the real one.

As

$$E^2 = \frac{1}{2p^2} - \frac{1}{2p'^2} = \frac{\sum Ri^2}{n} - \frac{\sum ri^2}{n} \text{ and } E = \frac{\sum Ri}{n}$$

Then

$$E = \frac{\sqrt{\sum ri^2}}{n\,(n-1)} \text{ if one accepts } \sum Ri \,.\, Rj = 0.$$

E is usually called standard error.

E decreases if the number of observations n increaes and if the precision (connected with ri^2) remains constant or increases. The correlation between E and Student's t is given in classical treatises [49].

The extension of the reasoning to qualitative observations is possible if a fixed scale can be described to classify the qualitative observations. To each observation one can then attribute a number which is its rank in the scale. A mean value can then be calculated and the standard error determined.

One could accept $\frac{m - E}{m}$ as definition of the plausibility of a set of n observations $m_1, m_2 \ldots m_n$ having a mean value m. This definition differs from the one suggested above in conjunction with the table of t values. A choice has to be made once and for all.

I do not say anything in general about the addition of plausibility E to the plausibility "that E is not true", and I shall admit: plausibility $E \leqslant$ objective probability of E to be true. I shall never use a subjective probability to determine plausibility. The value of a plausibility is always superior to a corresponding subjective probability.

The usual scale for probability is the closed interval (0, 1); 0, corresponds to a probability nil and 1 to certainty; $\frac{1}{2}$ corresponds to uncertainty. I shall however for plausibility use the scale (-1, 0, $+1$), where -1 means certainly false, 0 means uncertainty and $+1$ means certainly true. The advantage of such a scale is to avoid the necessity of considering both the opposite concepts or propositions in artificial memories; some demonstrations are also simplified by the use of the scale (-1, 0, $+1$).

2.2 FUNDAMENTAL THEORY

Minimum notations are necessary to designate: sets, propositions, relations, probabilities and plausibilities.

Sets are represented by capital letters.
Propositions are represented by small letters.
A term, which can be a *set or a proposition*, is represented by underlined capital letters.

p = *plausibility*; P = *probability*.

The *relations* we will consider are:

Correlation	C	$A\ C^p\ B$, meaning A is correlated to B with a plausibility p.
partial inclusion:	i	$A\ i^p\ B$, A is partially included in B with a plausibility p.
total inclusion:	I	$A\ I^p\ B$, A is included in B with a plausibility p.
partial implication:	$a\ i^p\ b$	a implies partially b with a plausibility p.
total implication:	$a\ I^p\ b$	a implies b with a plausibility p.

Equality is designated by =. It means possible substitution. This is all that need be considered for the time being.

From a purely mathematical point of view it would be simpler to consider plausible points defined as $(A, B, p)^i$ corresponding to Ai^pB, A, B and p being coordinates in the space of partial inclusion. But the handling of these points supposes a special training and the acceptance of a new way of thinking in logic. I suggest them for future consideration.

2.2.1 *Fundamental axioms and definitions*

The axioms I propose for the logic of the plausible must satisfy several criteria:

—They must lead to the rules observed in plausibility.
—They must not lead to irreducible contradictions.
—They must be considered as provisional and be perfected in practice.

Axiom 1. The plausibility of a statement should not be superior to *the best objective probability* that the statement should be true (on the same scale).
Axiom 2. The plausibility *based on scientific method* supersedes any plausibility based on subjective probability.
Axiom 3. A property valid for a relation R between two sets or proposition is valid for the relation R' if $R'IR$. (This is called a *meta property*).

Definition: The plausibility of a set, a proposition or a relation between two definite sets or propositions is equal to the weighted mean of all the plausibilities of the same.
Consequence: The weight of a scientific determination is infinite compared to the plausibility of a subjective appreciation.

Axiom 4. If A is included in B, then part of B is obviously included in A, and for plausible inclusions $A\ I^q\ B$ leads to $B\ I^q\ A$, q being inferior to p. In the practical cases I have used in logic, one could furthermore write:

$$q = \frac{\text{extension } A}{\text{extension } B}$$

Axiom 5. If a implies b or a I^p b, one can write by analogy:

$$a\ I^p\ b, \text{ then } b\ I^q\ a$$

$$\text{but } q \text{ becomes: } p.\ \frac{\text{intension } b}{\text{intension } a}$$

Transitivity. As there is a property of symmetry, even incomplete, it is interesting to find out if there is no property of transitivity in the relations of inclusion and implication.

If $A\ I^p\ B$ and $B\ I^r\ C$, what about $A\ I^x\ C$?

If we admit a homogeneous dilution of the extension of A in B and then of the extension of B in C, then $x = p.q$. It appears, once again, that this is the case in usual logic, the one we use, as confirmed by the examples given in "generalized logic". I may then admit:

Axiom 6. If $A\ I^p\ B$ and $B\ I^r\ C$, then $A\ I^{p \cdot r}\ c$, and similarly if a $I^p\ b$ and $b\ I^q\ c$ then $a\ I^{p \cdot q}\ c$.
Axioms 4, 5 and 6 are obviously valid *only* under the conditions of homogeneity described.
But if $P<0$ and $q<0$, then $A\ I^{-p}\ B$ and $B\ I^{-q}$ does not lead to $A\ I^{pq}\ C$.
One cannot write:

The blacks are not white.
The whites are not yellow.
Then the blacks are yellow.

Axiom 7. In the case of correlation C between two sets or two propositions: $A\ C^p\ B$ and a correlation between B and D, or $B\ C^q\ D$, a variation of A does necessarily affect D. Let us take p equal to the correlation coefficient existing between A and B or $s_{A.B}$:

If $A\ C^p\ B$ and $B\ C^q\ D$, then $A\ C^{k \cdot p \cdot q}\ C$
K is easy to calculate and is <1.

More generally axioms 5, 6 and 7 can be combined in:

If AR^pB, BR^qC, p and q being determinable, then $AR^{k \cdot pq}C$; $K \leqslant 1$, R being any relation.

Definition: If $-B$ means the negation of B
then $AR-B$ is equal to $AR^{-1}B$

I am now in possession of remarkable properties. They must of course be used with care, in the exact conditions in which they are valid. I may consider a chain of relations: $ARBC$ . . . and I may calculate by symmetry or transitivity the plausibilities of the relations in favourable cases. I may even close the chain into a circuit. I have a game which could become important if it were proved that any kind of reasoning can be presented in the form of a chain or a circuit. Let us play and try to determine some of the limits in which the game is valid. A simple case is given by a circuit of three relations.

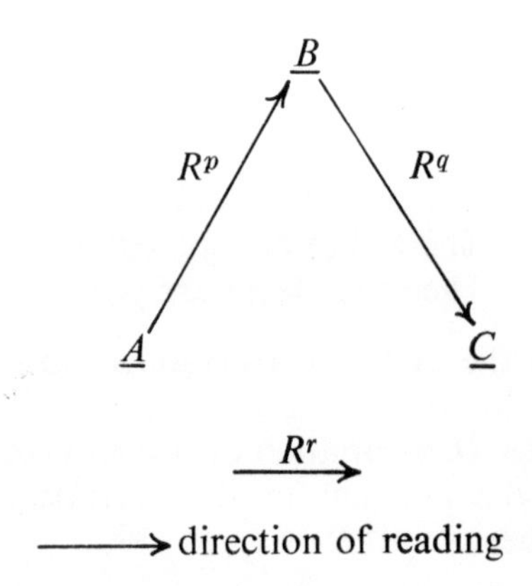

Axioms 5 and 6: $\left.\begin{array}{l} r = p.q, \\ p = r.q', \end{array}\right\} r = r.q.q'$ or $qq' = 1$

which can only be true for $qq' = 1$, q and q' being $\leqslant +1$. Therefore one can always write $r = p.q$, but the property cannot be true *at the same time* for all the sides of the triangle $A\ B\ C$ if p, q and r have well determined values. This is understandable: $r = p.q$ means that $r \leqslant$ to the smallest number p or q. The quality of being r is not shared necessarily by all the sides of the triangle except for the case where $p = q = r = 1$. This is a very important restriction which did not immediately spring to my mind. It is so easy to be wise afterwards.

2.2.2 *Theorem*

Let us consider a chain of identical relations, direct or symmetrical, having known plausibilities $p1$, $p2 \ldots pn$, all (or all but one) superior to 0.

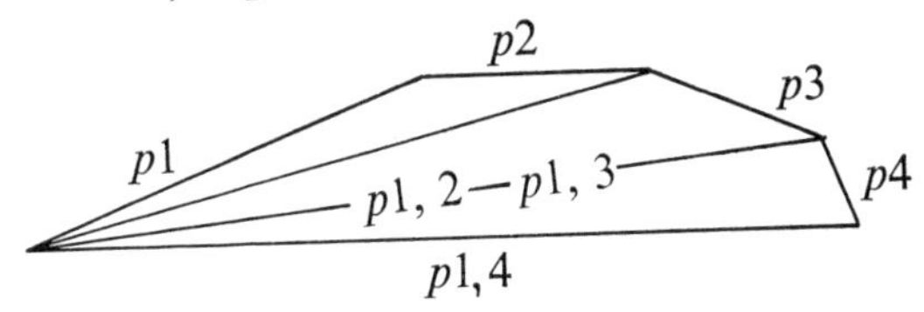

Axiom 7: $p1, 2 = K1.$

$p1, 3 = K1.K2.p1, 2.p3.$

$\ldots$

$p1, n = K1, K2 \ldots Kn.p1p2 \ldots pn;$

$K1, K2 \ldots Kn \leqslant 1$

$p1, n$ is the smallest value of p.

If $K1, K2, \ldots Kn = 1$, R is an inclusion or an implication. A general relation exists between all the plausibilities of a chain, if they are all taken in the same direction.

The passage from a chain to a circuit is in fact easy: $p1, n$ closes a chain into a circuit, but there is a privilege point in a normal circuit: point 1 in this case, which is called *origin*.

2.2.3 *My attitude*

My attitude is based on three principles:

1. Reasoning is improving continuously, especially with the increased complexity of the problems to be solved. One must try to extract from the experience acquired in reasoning as many systematic procedures as possible. The source of inspiration in logic lies in experience.
2. Mathematics provides us with a language which is at the same time simple and powerful: relations, symmetry, transitivity . . . One must introduce it into logic.
3. The passage from empiricism to formalism must be very progressive: reconstructions are inevitable.

I have on occasions made mistakes. I have more often expressed myself clumsily, but the aim has been constant.

2.3 GENERALIZED LOGIC

First of all I shall consider several *classical* forms of reasoning and try to write them in the form of a chain or a circuit to which the fundamental theorem applies. Inclusion and implication will be the only relations used in 2.3.

2.3.1 *A syllogism in Barbara*

White men are men (1)
All men are mortals (2)
White men are mortals (3)

Let us put:

White men = X; Y = all men; Z = mortals.
I = complete inclusion.

(1) becomes: $XI^{p}Y$ (here $p = 1$)
(2) becomes: $YI^{q}Z$ ($q = 1$)
(3) becomes: $XI^{r}Z$

Axiom 6, $r = p.q$ (here $r = 1$)
This is a very simple case.

2.3.2 *A syllogism in Camestres* [*51*]

No self-luminous bodies are planets (1)
All stars are self-luminous (2)
Therefore no stars are planets (3)

Let us say:

Self-luminous bodies = X
Planets = Y
Stars = Z

(1): $XI^{-p}Y$
(2): $ZI^{q}X$
(3): $ZI^{-r}Y$

Indeed, if I write the relations in another order so as to apply axiom 5:

$ZI^{q}X$, $XI^{-p}Y$ therefore: $ZI^{-qp}Y$.
$-r = -pq$. (axiom 6)

I should of course have replaced p and q by *1* in the example. By using p and q *I have generalized the syllogism into a plausible one*. I shall continue in the same manner.

2.3.3 *A syllogism in Boarko*

(1) All X's are Y's
(2) All Z's are not Y's
(3) Some Z's are not X's

(1) $XI^{p}Y$
(2) $ZI^{-q}Y$
(3) $ZI^{-r}X$

Indeed:

$ZI^{-q}Y$, $YI^{p'}X$ then $ZI^{-p'q}X$

$YI^{p'}X$ is the symmetrical relation of $XI^{p}X$ (axiom 4)
If we recollect that this kind of syllogism has been the source of a great deal of trouble [51] in the past, then it is amazing to see how simple they appear in the light of the new axioms. I should not, however, overestimate my achievement. Boole had already achieved the same result. All that I have added so far is the *plausible syllogisms* by replacing $+1$ or -1 by $+p$ or $-p$.

Let us now choose a *generalization* and try (and this is completely new) to deal with it by a calculus, as we did for syllogisms:

> While walking along the shore of Lake Geneva I observed that all the big white birds floating on it were swans and furthermore that they were white. Are all the swans white by any chance? So far I am not supposed to have observed any black swans. Let us put this in a mathematical form:
>
> $X =$ All the big birds on Lake Geneva
> $Y =$ All the swans
> $Z =$ All the white birds

(1) XIY
(2) XIZ

From (1) YI^pX (axiom 5) $p < 1$
YI^pX and XIZ, then $YI^{p\cdot}Z$ (axiom 7)

We have a typical plausible reasoning, the conclusion is not certain and p depends on the quotient $\frac{\text{extension } X}{\text{extension } Y}$ which is perfectly logical.
In order to correct the usual wrong reasoning which concludes from a limited number of observations that "all the swans are white", Reichenbach added a statement: "Seldom are all the animals of the same species at the same time the same colour." This may be represented by:

V = all the animals of the same species
W = beings of different colours

This leads to: VI^qW.
and YIV, VI^qW, then YI^qW.

Such a reasoning does not add anything to our first one; it is not required with plausible logic.

2.3.4 *Typical case of analogy*

If A and B both belong to the set C with respective plausibilities p and q, then the properties D of A can be extended to B with a plausibility r. In symbolic language:

$$A\ I^p\ C$$
$$B\ I^q\ C$$
$$A\ I\ D$$
$$B\ I^r\ D \qquad r\ ?$$

This leads to:

$B\ I^q\ C$

$C\ I^{p'}\ A$ (Axiom 5 applied to $A\ I^{p}\ C$)
$A\ I\ D$

$$\left.\begin{array}{l}\text{therefore: } B\ I^{qp'}\ D \\ \text{compared to } B\ I^{r}\ D\end{array}\right\} r = qp'$$

There is no difficulty in calculating the value of an analogy, conditions D being generally satisfied.

2.3.5 *A special case of generalization is extrapolation*

If B is an extrapolation of A, this means that A and B are both points of a set C which contains all values $A, B \ldots$; therefore:

$$\begin{array}{l} A\ I\ \ C \\ B\ I^{p}\ C \end{array}$$

I observe for A certain properties D; $A\ I\ D$. There is a plausibility that this property will also be observed for B, or $B\ I^{q}\ D$. What is the value of q? Let us summarize the relations and the question:

$$\begin{array}{ll} A\ I\ C & \\ B\ I^{p}\ C & \\ A\ I\ D & \\ A\ I^{q}\ D & q? \end{array}$$

This leads to:

$B\ I^{p}\ C$
$C\ I^{r}\ A$ (Axiom 5 applied to $A\ I\ C$)
$A\ I\ D$
or $B\ I^{p.r}\ D$ (Axiom 7 applied to the three preceding relations)

We already have $B\ I^{q}\ D$, therefore $q = p.r$.
The nature of this extrapolation lies between analogy and

generalization, the generalization being the passage from A to B, the analogy being the extension of properties D because A and B belong to the same set C.
More complex chains or circuits can easily be considered as I have done. As far as can be judged I have created A GENERALIZED LOGIC.

2.4 INDUCTION

Have I solved all the problems of logic by introducing the logic of the plausible? Let us consider the difficult case of induction [52]. This all depends on the meaning given to the term "induction." If one means generalization, I have genuinely progressed. If one means the effect on a statement of a collection of similar results, no improvement is apparent. The factors to be considered are:

Convergence of observations or reasonings.
Degree of independence of observations or reasonings.
Precision of observations or *plausibility* of reasoning.
Number of observations and reasonings.

I am not considering the classical part of probability which deals with collection of data, the value of the mean, weights . . . as I do not have anything to offer as an improvement in this well developed area. But we should at least know how to combine the elements of the new logic, in other words we should define a few essential operations.
I have used three main *operations; union, intersection* and *abstraction.* A union of identical *relations* orientated the same way and having the meaning "A or B" is the relation concerned having as first term the union of its first terms, as second term the union of its second terms, and as

plausibility a number comprised between the lowest (in absolute value) of the plausibilities and the mean plausibilities of all the relations. The smallest value is chosen if the terms are independent, the mean value is chosen if the terms are identical, but a union having the meaning "*sum* $A + B$" leads to $p.\ (A + B) = pA + pB$.
An *intersection* of identical relations orientated the same way is the relation concerned having the intersection of the first terms as first term and the intersection of the second terms as second term, the plausibility of the relation being the lowest plausibility.
An intersection is called an *abstraction* when its plausibility is the higher of the two.
A comparison with probabilities is necessary:

$P\,(A + B) = PA + PB$ if $A\,.\,B = 0$ which is a sum.
$P\,(A\,.\,B\,) = PA\,.\,P\,(B : A)$
$P\,(B : A) =$ probability of B if A is observed.

This seems simple enough but there is a danger in the use of scales: the formula $P\,(A + B)$ is valid only if the scale (0, 1) is used. The same applies to $p\,(A+ B) = pA + pB$. The use of scale (-1, 0, $+1$) has many advantages in the logic of the plausible; it becomes difficult in the case of $p\,(A + B)$ and $p\,(A + B)$.
So far I have added nothing to the theory of induction, but we are better equipped for dealing with simple combinations of observations or reasonings.
I am of the opinion that if several independent reasonings converge to the same conclusion, the plausibility of the conclusion should be superior to any of the separate plausibilities, but of course <1 in absolute value. I also feel that this should not be true if the separate plausibilities are too small in absolute value. A great number of small reasons does not necessarily constitute a proof.

$$\left.\begin{array}{l} b\ i^{p}\ a \\ c\ i^{q}\ a \\ \ldots \\ n\ i^{r}\ a \\ b.c\ldots.n = 0 \end{array}\right\}$$

Let me represent *convergence by* $+$, sign already used for addition:

$$(b + c + \ldots + n)\ i^{x}\ a.$$

What is the value of x? I have no way of calculating x from a previous axiom or operation. I must, therefore, invent something new for our satisfaction:

- x must be superior to $p, q, \ldots r$.
- x must be < 1 in absolute value.
- x must increase with the values of $p, q \ldots$ and with their number.
- x must be reduced to p if one reasoning is made.
- x must be such that for a great number of small plausibilities it never approaches unity.
  or else x should only be considered for plausibilities high enough in absolute value (0·7 or above).

This expresses one aspect of induction for which I find no logical foundation. Let us call x: *plausibility of convergent and independent high plausible reasonings.* One may reject or accept this point of view which I am putting forward for consideration.

I started from a psychological attitude aiming at a logical rule. This is the historical origin of logic from Aristotle onwards.

Chapter Three

INTRODUCTION OF SOME IMPORTANT APPLICATIONS

The new logic described in Chapter 2 leads to progress in many fields. I shall here be considering several of them as an introduction to a possible more profound study. I have not mentioned forecasting as a possible field of application. Forecasting is intimately connected with plausible reasoning and creativity, so I have not thought it necessary to stress it once again.

3.1 SCIENTIFIC REASONING

If important scientific reasonings are transcribed into symbolic language and if generalized logic is used, the general conclusion is that:

they are often very *simple*
and most of the time *unexpected.*

They could probably have been found earlier if *a systematic approach* and the right symbolism had been used from the start.

The Curies' reasoning in the discovery of radium can be summarized as follows:

Pitchblende is a mineral (1)
Pitchblende is radioactive (2)
Other minerals could be radioactive (3)

This can be transcribed in:

(1) Pitch. R Minerals
(2) Pitch. R^p Radioactive bodies
(3) Minerals R^q Radioactive bodies

(1) Pitch. R .Min. gives Min. R^r Pech. r is small, but > 0.
(2) Pitch. R^p. Radioactive bodies.

Axiom 7: Min. $R^{r.p.}$ Radioactive bodies.
$q = rp.$ q is positive.

By playing with the data and considering their plausibilities such a reasoning was inevitable.

Let us now consider the structure of the reasoning which leads to de Broglie's hypothesis.
The corpuscular theory of light (*tc*) explains the photoelectric effect (ph.) but not diffraction.
$tc\ .\ i^{p}\cdot ph$ (1); to . i^q di (2)
The undulatory theory of light (*tu*) explains diffraction (*di*); but not the photoelectric effect.
to i^r *di.* (3); $tc\,.\,i^{-}s\,.\,ph.$ (4)
Both the photoelectric effect and diffraction have been observed,
$obs\,.\,i^t\,.\,ph$ (5)
$obs\,.\,i^u\,.\,di$ (6)

Let us write the preceding equations in a chain:

$$tc - i^{p} - ph - i^{t'}$$
$$i^{-q} \qquad obs$$
$$to - i^{r} - di - i^{u}$$

I have replaced (5) and (6) by two symmetrical relations. A circuit could be closed if we connect *tc* and *tu* to a more general theory: *tg* from which *tc* and *to* can be derived.

$$tg\ i^{v}\ tc$$
$$tg\ i^{w}\ tu$$

Therefore:

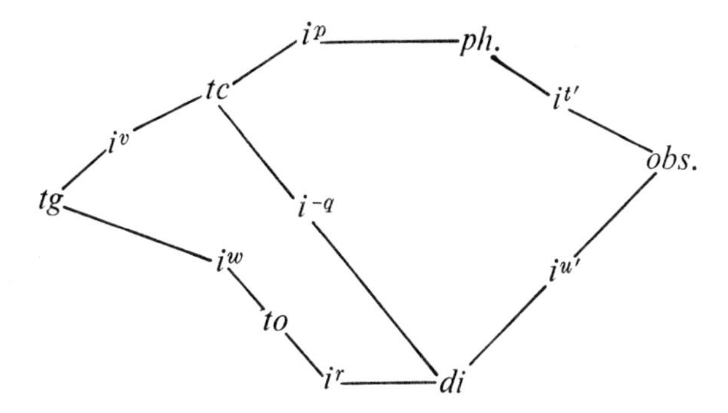

tg is the hypothesis de Broglie aimed at in his theory.

3.1.1 *Anomaly*

Anomaly is one of the most usual sources of new theories [53]. A good example is the discovery of adsorption by Langmuir [54].

It was thought that the volume of gas absorbed by a

porous metal was rather small (1). Langmuir observed a release of gas incompatible with those thoughts (2). He explained it by a new theory called *adsorption* (3).

theory 1 i^p observed facts.
theory 1 i^{-q} new observed facts.

One must find a theory 2 to explain the new facts. The situation is similar to the one described in 3.1.2.

I have not of course described all the forms of scientific reasoning. Some of these reasonings are analytical, others are combinatory. They are seldom complex. One of the best examples of analysis combined with axiomatics is described in Newton's *Principia*. IT APPEARS THAT A GOOD SYMBOLISM AND GENERALIZED LOGIC IS A NEW IMPORTANT INSTRUMENT IN THIS AREA.

3.2 PLAUSIBILITY OF SCIENTIFIC HYPOTHESES

What is the plausibility of an hypothesis formulated *from observations*. Let us assume that a, b and c are independent observations which lead to a phenomenon f. Independent is taken here in its strongest meaning.

If a, then f, with an objective probability Pa and a plausibility pa.
If b, then f, with an objective probability Pb and a plausibility pb.
If c then f, with an objective probability Pc and a plausibility pc.
If $a + b + c$ (sum), then f with a probability $Pa +$

$Pb + Pc$ and a plausibility $pa + pb + pc$, in the scale (0, 1).

A step further is the *formulation of a theory* by analogy in order to explain observed facts.
"The pressure observed in a vessel (v) may result from "the impact of small particles" (c) or "an internal force" (f) or also from "a gas contained in the vessel" (g). There is a plausibility x that "a gas contained in a vessel" is "composed of particles" (h). The equations are:

$$\left\{\begin{array}{l} c\ i^p\ v \\ f\ i^q\ v \\ g\ i^r\ v \end{array}\right.$$
$$x\ ?$$

$g\ i^r\ v,\ v\ i^{p'}\ c$ (Axiom 5)
then $g\ i^{r \cdot p'}\ c. = h.,\ r\ p'$ is small but positive.

The value of an hypothesis will vary with the nature of the verifications made. This problem has been studied by George Polya [34] who proposed a series of theorems. He unfortunately proved them by using, even with subjective probabilities, the property I have underlined as being valid for objective probabilities only, that it is to say: $pA + p\bar{A} = 1$. I shall try to prove Polya's theorems without making the same mistake.
Theorem 1. The verification of a consequence of an hypothesis H increases the plausibility of that hypothesis, and the increase in plausibility is bigger the less plausible was the fact verified.

$E(b)$ = set of the points or elements which are the logical consequences of H, H being the hypothesis formulated.

$E(b) . I^p\ H.$ (1)

$E'(b)$ = sub-set of $E(b)$ which has been verified.
$E'(b).I^q\ E(b)$ and $E'(b).I^r\ E(t)$
$H.I^x\ E(t) \quad x$?

One may write $H.I^{p'}\ E(b)$; $E(b)\ I^{q'}\ E'(b)$; $E'(b)\ I^r\ E(t)$ or $H.I^{p'.q'.r}\ E(t)$
If r increases, $p'.q'.r$ increases. If r were small the increase $p'.q'.r$ would become bigger.

George Polya enunciated the same property as follows: "The verification of a consequence renders a conjecture more credible" and "The increase of our confidence in a conjecture due to the verification of one of its consequences varies inversely as the credibility of the consequence before such_ verification". In his demonstration he replaces $Pr\ (A)$ by $\{1 - Pr\ A\}$ [35, II, p. 121] making use, therefore, of a relation valid only for objective probabilities.

Theorem 2. The observation of facts analogous to an hypothesis H increases the plausibility of H.

$E(c)$ = set of elements or points analogous to H.
$E(b)$ = set of elements or points which characterize H.
$E'(c)$ = sub-set of $E(c)$ which belong also to $E(b)$.
$E(t)$ = set of true facts.

$$\begin{cases} E'(c)\ I^p.E(b) & \\ E'(c)\ I\ E(c) & \text{and } E(b)\ I\ H \\ E(c)\ I.^q\ E(t) & \end{cases} \quad p \text{ have a high value.}$$

$H.I^x.E(t)$; x ?

$$\left.\begin{array}{l} H.I^r\ E(b)\ \text{(Axiom 5)} \\ E(b)\ I^{p'}\ E'(c) \\ E'(c)\ I.E(c) \\ E(c)\ I^q\ E(t) \end{array}\right\}\ H.I^{r.p'.q}\ E(t)$$

If p increases, p' increases and $r.p'q$ increases, proving theorem 2.
Theorem 3 was not enunciated by George Polya, as far as I know, but is analogous to the ones discovered by Polya.
Theorem 3. The observation of a consequence contrary to a hypothesis H decreases the plausibility of H (even reduces it to a negative value).

H = hypothesis
$E(-b)$ = set of consequences contrary to H.
$E'(-b)$ = sub-set of $E(-b)$
$T(t)$ = set of true facts.

$$\begin{cases} E'(-b)\ I\ E(-b) \\ E'(-b)\ I^p\ E(t) \\ \qquad p \text{ has a high value} \\ H.I^{-q}.E(-b) \end{cases}$$

then:

$$H\ I^{-q}\ E(-b),\ E(-b)\ I^r.E'(-b),\ E'(b).I^p\ E(t)$$

therefore:

$$H.I^{-q.r.p.}\ E(t) \text{ which proves theorem 3.}$$

George Polya expressed another property the following way:
Theorem 4. "The more confidence we place in a possible ground of our conjecture, the greater will be the loss of faith in our conjecture when that possible ground is refuted". His reasoning, once again, uses the property $pA + p\bar{A} = 1$ [35, II, p. 123].
This is easily proved otherwise:

G = possible ground for conjecture H

$$H\ I^p\ G$$

$G =$ has a plausibility q not to be true

$$G\ I^{-q}\ E(t)$$

Therefore

$$H\ I^{-pq}\ E(t)$$

All the theorems of George Polya and others can be proved by using the new theory without any confusion between objective and subjective probabilities.

3.3 A SYSTEMATIC METHOD OF CREATIVITY

Methods of creativity, as they are called, are often a mixture of psychological and logical procedures aimed at improving the discovery of new ideas. They try to reduce polarization of the brain on accepted thoughts, so allowing the birth of new ones; they also try to enrich the number of associations of ideas by varied recipes: some of them go as far as trying to excite the brain in the expectation of increasing its power.
I propose the following rough classification of these methods from the point of view of the kind of procedures used:

The logical methods

analysis
synthesis
comparison
simplification
supposition
substitution
iteration

can be combined between them into:

dialectic

generalized logic
axiomatics
classification
experimental method
algebra of solution

or used directly or indirectly:

straight thinking
lateral thinking

The psychological methods

brainstorming
synectics
semi-dream and hypnotic methods
choice of inventors and creators

can be used in combination with the preceding logical methods.

It is obvious to me that complete freedom of thought must be advocated in art where the aim is to create emotions, but that freedom is automatically restricted in fields where we aim at discovering the laws and constitution of nature and at constructing new instruments which achieve practical purposes. Of the latter cases systematic thinking (*not to be confused with conventional thinking*), the only one I am considering in this book, becomes more important. As an example I shall consider a case in which brainstorming was used in comparison with generalized logic.

The problem was to find a new kind of bread. The factors considered were:

composition
form
density

taste
usage
price
preparation

This may result from a morphological analysis but was in fact obtained after a brainstorming session. This already leads to a great number of possibilities which may be combined in many possible solutions. But are we sure that the list is complete? Let us start again from several of the preceding words, for instance taste. Taste is included in the sense-organs.

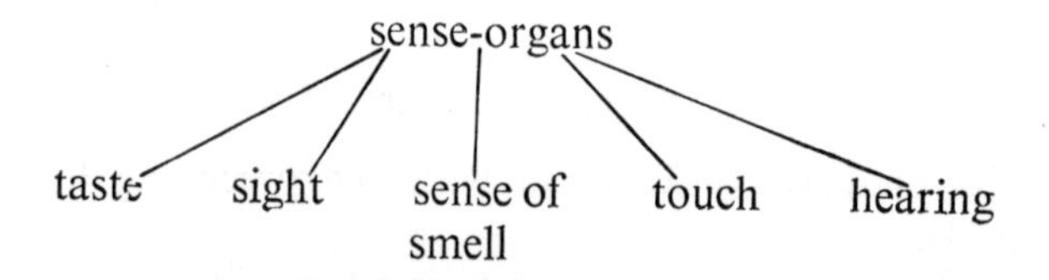

I am surprised that I have left out:

the appearance of the bread
its odour
the sensation one has in touching it
the crackling sound when it is pressed

which are all important factors (at least to the French). Analogy has achieved an important aim in extending an analysis which previously appeared as complete. From odour another analogy could lead to new ideas. Generalization could be used in a similar way. The result obtained by systematic analysis was far richer than the one obtained by brainstorming.

The result of a complete analysis is to make the number of combinations rapidly into a very high number. In order to

reduce that number it is suggested to calculate the plausibilities of the factors combined so as to get an idea of plausibility of the combination and to choose the combination for which the plausibilities are within an accepted range.
I suggest the new method be termed: *plausible exploration.*

3.4 HEURISTICS, AN APPROACH TO A NEW SCIENCE

The possibility of existence of a science of invention and discovery has often been discussed. In a thesis entitled "Logic of discovery" presented by B. G. Buchanan (Michigan State University, Philosophy, 1966), the author's conclusions are rather pessimistic. It is obvious that a plausible logic could change the position in an important way by the combination of what has previously been found by the researchers in that field together with my new approach.
Creating a theory is building a *system* which coordinates all the known data into a structure which is as complete and as simple as possible, the ideal being the axiomatic form creating a new instrument organizing the known properties and the ones which have a high plausibility of being true into a *system* which solves the problem as efficiently as possible. In both cases creating is *building a system.*

3.4.1 *The laws of scientific systems*

Knowledge or instruments, are the following:

Principle One: the content of information of a closed

system remains constant. The content (energy + matter) of a closed material system remains constant.

Principle Two: the mean plausibility of a closed system of information tends to increase by the use of reasoning. Entropy in a closed material system increases with time.

Principle Three: for a given content of information in the case of an axiomatic system one cannot reduce the number of axioms under a certain minimum. The use of plausibility may decrease the number of those axioms. The entropy of a perfect crystalline substance is nil at zero degree absolute.

3.4.2 *The principles of heuristics*

Principle One (*richness*): one must collect as many data and formulate as many reasonings as possible.

Principle Two (*economy*): a maximum result must be obtained by the least possible use of energy. The conflict between principles one and two is obvious. A compromise is necessary.

Principle Three: to answer the questions continually or occasionally asked the results must be adequate.

Principle Four: the final aim of invention or discovery is for the improvement of the quality of life.

Heuristics becomes the application of the four principles just enunciated to the creation of systems as defined by their own three principles. This presents the situation from a new angle. The interest of generalized logic is to make possible the application of the axiomatic form of systems to all cases by replacement of the rigorous classical logic by the logic of the plausible. This is an important step.

3.5 LATERAL THINKING

Is lateral thinking, advocated by de Bono [55], made easier by the use of generalized logic? I have chosen the surprising reasoning made by the teenage girl in the pebble story proposed by de Bono as a typical example of lateral thinking.

Faced with the choice between two pebbles, one meant to be black and the other white, dropped into an empty money-bag, the girl observes that the money-lender has cheated by putting two black pebbles in the bag. The choice could have frightful consequences. A black pebble would mean marrying the old and ugly money-lender and her father's debt being cancelled; a white pebble would mean freedom for the girl and her father; should she refuse to choose, her father would go to jail. Let us use a minimum of symbolism:

choice of white pebble	= wh.
choice of black pebble	= bl.
refusal of choice	= ref.
freedom for the girl and her father	= fr.
marrying the money-lender	= mar.
father in prison	= pr.

In a normal course of events the possible cases, including errors or cheating, are:

Initial situation	*Action*	*Consequences*	
no pebble	→ ref.	→ pr.	(1)
two black pebbles	→ bl.	→ *mar.*	(2)
two white pebbles	→ wh.	→ fr.	(3)
two different pebbles	→ wh.	→ fr.	(4)
	→ bl.	→ mar.	(5)

The teenage girl is faced with case (2). Let us extend the reasoning by adding the colour of the pebble remaining after the choice and the nature of the situation, right R or wrong W:

Initial situation	*Action*	*Final situation*	*Consequence*	
Ⓡ no pebble	→ ref.	→ no pebble remains	→ pr	(1)
W two black pebbles	→ bl.	→ black remains	→ mar	(2)
W two white pebbles	→ wh.	→ wh.	→ fr.	(3)
Ⓡ two different pebbles	→ wh.	→ black remains	→ fr.	(4)
	→ bl.	→ white remains	→ mar.	(5)

As the case should be Ⓡ and if the pebble remaining is black, then the result is freedom. It is the logical choice for the teenage girl. To reach it she drops the black pebble she has taken from the bag and suggests considering the remaining pebble which is also black. Under the right conditions the conclusion is given by (4).

Lateral thinking is a reasoning extended far enough so as to include all the essential aspects of the reasoning made. It is not a mystery.

3.6 DOCUMENTATION

It is quite unnecessary to remind the reader of the growth of documentation [56]. It is a phenomenon well known among scientists. So far no solution has been found. I do not of course pretend to have solved the difficulty. I nevertheless hope to propound an interesting approach. It presupposes the analysis of any *interesting* and *original* document by a person familiar with the logic of the plausible and who would be able to analyse the papers produced. This constitutes a real effort of time and expenditure which has sometimes been considered as prohibitive. The normal way of decreasing the cost of such an undertaking is to distribute the work between the various scientists working in the same centre.

I am stressing the fact that any reasoning can be decomposed into couples of relations which is the basis of my logic, and that relations can be combined according to known rules or operations which are given in my logic. This is all I require for proposing a new method of documentation. Operations on couples may be done by a computer. The analysis must be intelligently carried out by men who understand perfectly the structure of the reasoning. I found in trying the process that it often happens that some essential ideas are not explicit in the text, even in those which are more precise [57]. A complete reasoning must be presented in the analysis.

Analysing into couples, recombining all the couples with the same members and relations but different plausibilities, lead to a new series of couples which condense all the knowledge acquired into the shortest possible form. One could even go further and organize the couples into new circuits, use analogy, generalization . . . and reason scientifically on a more certain basis. This is my reason for proposing that the system be named "Euredoc", short for heuristical documentation.

As is customary in documentation a careful choice of vocabulary for constituting a *thesaurus* should be made and the right level of analysis selected. The passage from a level of analysis to another is feasible, but often leads to some complication.

Other than the use of computer already mentioned, it is possible to use the same method in a simple mechanical manner. One may, for instance, represent each concept or proposition by one vertical and one horizontal line which intersect at the point representing the concept or proposition. One point, and one only, is used for a concept or proposition. The intersection between a horizontal line belonging to concept A and a vertical line representing B corresponds to:

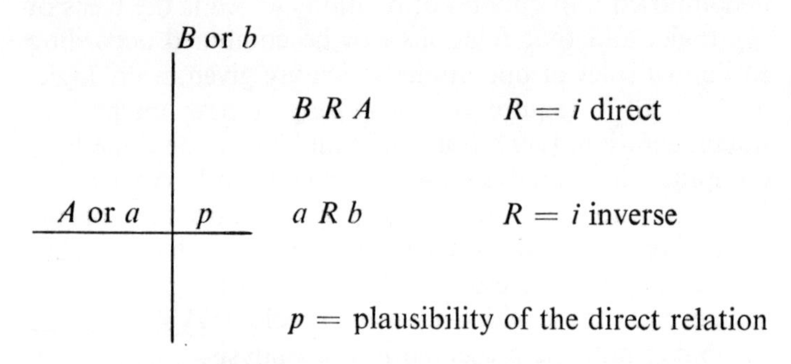

The form of the point varies with the nature of the relation, the value of p varying with the results obtained.

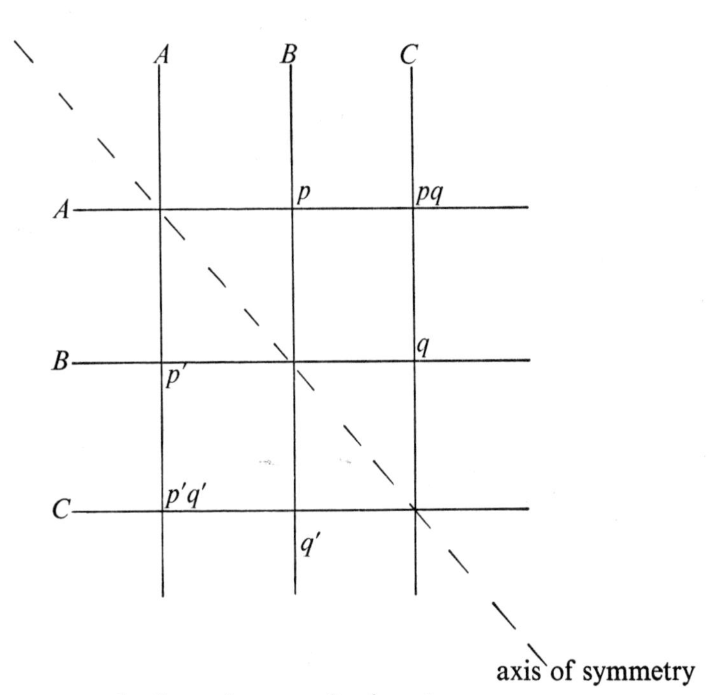

$p\,.\,q$ results from the use of axiom 1.
p', q', $p'q'$ result from the use of axiom 5.
Remark. If the lines are deleted and the plausibilities only are written in their respective order one obtains for concepts A, B, C:

$$\begin{vmatrix} \mathrm{I} & {}_{,}d & {}_{,}b{}_{,}d \\ b & \mathrm{I} & {}_{,}d \\ bd & d & \mathrm{I} \end{vmatrix}$$

which *can be used as a pseudo-matrix, a kind of generalization of the tables of truth.* Such pseudo-matrices have some interesting properties which can be observed when a line or column is displaced. The union or intersection of pseudo-matrices may also be considered. They are different from the properties of normal matrices.

One could easily establish the following property: If n plausible, identical relations form a chain, the total number of relations which may exist between all the concepts or propositions connected does not exceed n^2. I suggest this as a simple exercise for the reader's skill. This conclusion is important. The number of possible different relations is not infinite.

3.7 MEDICAL DIAGNOSIS AND JUDICIAL PROOF

The use of probability for making a diagnosis has been advocated by Professor W. I. Card of the University of Glasgow [58]. Diseases are defined by a number of symptoms. The problem of diagnosis becomes one of identification of a series of collected data with a list of symptoms, the value of each symptom being defined by the frequency of its occurrence in the disease. Or better still, if there should be a list of well established figures such that the existence of symptom 1, corresponds to a probability $P1$, to suffer from sickness $s1$, probability $P2$ . . . suffer from sS . . .

	sickness 1	sickness 2	sickness 3	
Symptom 1	$P1$	—	$P2$	. . .
Symptom 2	$P3$	$P4$	—	. . .
Symptom 3				
. . .				
Total				

there is no advantage in using plausibilities. If the list should be imperfect or incomplete, plausibilities may be used. The theorems of George Polya give an estimate of a hypothesis (here a given sickness) with the observation of symptoms, analogies . . . but such a study would need very careful development.

George Polya has stressed the interest of plausibility in judicial proofs. Each proof has a probability P to correspond to a fact F. The existence of the fingerprints of individual D on a gun means that D has touched the gun. The presence near a corpse of an object having belonged to the same individual D corresponds to a probability, subjective in this case, of individual D having been near the corpse. A motive for D to have an interest in getting rid of the person killed constitutes another circumstantial evidence in favour of D being the killer. Convergence of all these proofs increases the plausibility that D was the culprit. If the examination of the bullet found in the corpse leads to the identification, with a high positive probability, of the gun used by the killer and if that gun is the one touched by D, the probability of D being the culprit is further increased. D will then most probably be arrested and charged. The mechanism is simple.

Data d_1 leads to conclusion c_1 with plausibility p_1

$$
\begin{array}{l}
d_1 \; I^{p1} \; c_1 \\
d_2 \; I^{p2} \; c_2 \\
\ldots \\
\hline
d_1 + d_2 + \ldots \; I^{p} \; c
\end{array}
$$

The complete field of facts comprises:

the culprit chooses a gun
the culprit touches the gun

the culprit has a motive to kill
the culprit fires the gun
the gun kills the man found dead
. . .

One must compare the field of data obtained with the complete field and calculate the plausibility of the data obtained in order to determine the likelihood of culpability.

REFERENCES

1. M. S. F. Hood. "The tartara tablets". *Scientific American*, **218**, n°5, 30–37 (1968).
2. Thureau-Dangin. *Textes mathématiques babyloniens*. Leiden. 1938.
3. R. J. C. Atkinson. *Stonehenge*. Hamish Hamilton. London.
   G. S. Hawkins and J. B. White. *Stonehenge Decoded*. Souvenir Press. London. 1966.
4. Claude Lévi-Strauss. *La pensée sauvage*. Plon. Paris. 1962.
5. Paul Tannery. *Pour l'histoire de la science Hellène*. Alcan. Paris. 1887.
6. G. S. Kirk and J. E. Raven. *The Presocratic Philosphers*. Cambridge University Press. 1964.
7. E. Dupreel. *La légende socratique et les sources de Platon*. Brussels. 1922.
8. V. de Magalhaès-Vilhena. *Socrate et la légende platonicienne*. P.U.F. 1952.
9. B. Jowett. *The Dialogues of Plato*. Clarendon Press. Oxford. 1875.
10. François Lasserre. *The Birth of Mathematics in the Age of Plato*. Meridian books. Cleveland. 1966.
11. W. Jaeger. *Aristotle. Fundamentals and History of his Development*. Oxford University Press. 1934.

12. A. Wartelle. *Inventaire des manuscripts grecs d'Aristote et de ses commentaires*. Les Belles Lettres. Paris. 1963.
13. W. and M. Kneale. *The Development of Logic*. Clarendon Press. Oxford. 1962.
14. *Theophrastus' Enquiry into Plants* (translation by Sir Arthur Hort). Harvard University Press. 1916.
15. M. Clagett. *Archimedes in the Middle Ages*. University of Wisconsin Press. Madison. 1964.
16. J. L. Heiberg. "Eine neue Archimedes Handschrift". Reprint from "Hermes". Berlin. 1907.
17. Thomas L. Heath. *The Works of Archimedes*. Cambridge. 1897. P. Ver Eecke. *Les oeuvres complètes d'Archimède*. Vaillant Carmanne. 1960.
18. Vasco Ronchi. *The Nature of Light—an Historical Survey*. Heinemann. London. 1970.
19. M. Berthelot. *Introduction à l'étude de la chimie des anciens et du moyen-âge*. Librairie des arts et des sciences. Paris. 1900 (2nd ed.).
20 Martin Gardner. *Logic Machines and Diagrams*. McGraw Hill. Mossen Joan Avingo. *El terciari Frencesca Beat Ramon Lull*. Ignalda. Barcelona. 1912 (with a list of manuscripts).
21. J. G. Crowther. *Francis Bacon*. The first statement of science. The Cresset Press. London. 1960.
22. *The Works of Lord Bacon*. Reeves and Turner. London. 1889.
23. Descartes. *Oeuvres et lettres*. La Pléiade. Paris. 1953.
24. *La géométrie de René Descartes*. Hermann. Paris. 1927.
25. *Oeuvres de Fermat*. Gauthier-Villars. Paris. 1891.
26. Pascal. *Oeuvres complètes*. La Pléiade. Paris. 1954.
27. F. N. Davies. *Games, Gods and Gambling*. Griffin. London. 1962.
28. *Leibniz Selections* edited by Ph. P. Wiener—Charles Scribner's Son. New York. 1951.
29. *Leibniz Logical Papers*. A selection translated and edited by G. H. R. Parkinson. Clarendon Press. Oxford. 1966.
30. J. M. Child. *The Early Mathematical Manuscripts of Leibniz*. The Open Court Publishing Co. London. 1920.
31. E-N. da C. Andrade. *Sir Isaac Newton*. Auction Book. New York. 1958.
Louis Tranchard More. *Isaac Newton—A Biography*. Dover Publications. New York. 1934.
32. John Herivel. *The Background of Newton's Principia*. Clarendon Press. Oxford. 1965.

33. Henry G. van Leeuwen. *The Problem of Certainty in English Thought*. 1630–90.
Barbara J. Shapiro. "Law and Science in seventeenth-century England". *Stanford Law Review*, **21**, 727–63 (1969).
34. G. Polya. *Mathematics and Plausible Reasoning*. Princeton University Press. Princeton. 1954)
G. Polya. *Patterns of Plausible Inference*. Princeton University Press. Princeton. 1954.
35. M. Merriman. *Method of Least Squares*. New York. 1884. Transactions of Connecticut Academy, IV, 151. (1877).
D. E. Smith. *History of Mathematics*. Dover Publication. New York. 1951.
36. Sadi Carnot. *Réflexions sur la puissance motrice du feu*. New edition by Blanchard. Paris. 1953.
37. Ch. Rémond. *Les trois républiques et les trois Carnot*. Maurice. Paris. 1889.
Oeuvres Complètes de François Arago, edited by J. A. Barral. Volume I. Baudry. Paris. 1854.
38. R. Bourgne and J. P. Azra. *Ecrits et mémoires mathématiques d'Evariste Galois*. Gauthier-Villars. Paris. 1962.
39. George Boole. *An Investigation of the Laws of Thought on which are founded the Mathematical Theories of Logic and Probabilities*. Dover Publications. New York.
40. F. Yates. *Experimental Design* (selected papers). Charles Griffin. London. 1970.
41. "Ronald Aylmer Fisher (1890–1962)". *Bibliographical memoires of Fellows of the Royal Society*. Volume 9. November 1963.
42. Ph. Frank. *Einstein, His Life and Times*. A. Knoff. New York. 1947.
A. S. Eddington. *The Fundamental Theory*. Cambridge University Press. 1948.
43. Kurt Gödel. *On Formally Undecidable Propositions of Principia Mathematica and Related Systems*, translated by B. Meltzer. Oliver and Boyd. Edinburgh. 1962.
E. Nagel and James R. Newman. *Gödel's Proof*. New York University Press. 1958.
44. F. Zwicky. *Morphology of Propulsive Power*. Pasadena, California. 1962.
45. R. Leclercq. "The use of generalized logic in forecasting". *Technological Forecasting and Social Change*, **2**, 189–94 (1970).

Alex F. Osborn. *Applied Imagination.* Charles Scribner's Sons. New York. 1953.

46. W. Gordon. *The Metaphorical Way of Learning and Teaching.* Porpoise Book. Cambridge. 1966.
Edward de Bono. *The Use of Lateral Thinking.* Jonathan Cape. 1967.

47. A. N. Kolmogorov. *Grundbegriffe des Wahrscheinlichkeitrechnung.* 1933.
E. Borel. *La valeur pratique et la philosophie des probabilités.* Gauthier-Villars. Paris. 1939.
R. Carnap. *The Two Concepts of Probability in "Philosophy and Phenomenological Research*". 1945.
W. Edwards. "Behavioural decision theory". *Ann Rev. Psych.*, volume 12, pp. 437–98, 1961.
W. Edwards and A. Tversky. *Decision Making.* Penguin Books. 1967.
B. de Finetti. "Les problèmes psychologiques sur les probabilités subjectives". *Journal de psychologie normale et pathologique*, pp. 253–59, October–December 1954.
I. J. Good. *Probability and Weighing of Evidence.* Griffin. London. 1950.
M. Keynes. *Treatise on Probability.* 1921.
R. D. Luce and H. Raiffa. *Games and Decisions.* Introduction and critical survey. Wiley. New York. 1957.
F. P. Ramsey. *The Foundations of Mathematics and Other Logical Essays.* Harcourt Brace. New York. 1931.
L. J. Savage. *The Foundations of Statistics.* Wiley. New York. 1954.
R. von Mises. *Warscheinlichkeit, Statistik und Wahreit.* Springverlag. 1928.

48. Student. *Biometrika*, volume 6, pp. 1–25, 1908.
Fisher and Yates. *Statistical Tables for Biological, Agricultural and Medical Research.* Oliver and Boyd. London. 1949.

49. H. Jeffreys. *Theory of Probability.* Clarendon Press. Oxford. 1939. p. 118.

50. J. G. Kemeny and J. L. Snell. *Finite Markov Chains.* Van Nostrand. Princeton. 1960.

51. W. S. Jevons. *Elementary Lessons in Logic.* MacMillan. London. 1965. p. 146 and fol.

52. J. Patrick Day. *Inductive Probability.* Routledge and Kegan Paul. London. 1961.

L. J. Cohen. *The Implication of Induction.* Methuen and Co. London. 1970.
M. Boudot. *Logique inductive et probabilité.* A. Colin. 1972.
53. J. H. Hildebrand. "Order from chaos". *Science*, **150,** n° 3695, 1965, p. 441–49.
54. A. Rosenfeld, J. H. Hildebrand, E. Rideal and P. W. Bridgman. *Langmuir, The Man and the Scientist.* Pergamon. Oxford. 1962.
55. E. de Bono. *The Use of Lateral Thinking.* Pelican Book. London. 1971.
56. Derek J. de Solla Price. *Science since Babylon.* Yale University Press. 1961.
57. A. Einstein. *Annalen der physik*, **17**, 1905.
58. W. I. Card. *Logique nouvelle en médecine.* Science. 1971.

INDEX

Alexander, 6, 7
Alhazen, 9, 10
Analogy, 47, 48, 49, 57
Anomaly, 55
Antipater, 7
Archimedes, 8, 19
Aristotle, 5, 6, 8
Axioms, fundamental, 40, 41

Bacon, Francis, 13, 14, 21
Barrow, Isaac, 20
Binomial theorem, 19
Boole, George, 27, 28
Bravais, A., 26
Buchanan, B. G., 63

Cardano, 17
Carnot, N. L. S., 24, 25
Correlation, 26, 39
Creativity, methods of, 60–62
Curies, and discovery of radium, 54

De Bono, E.F.C.P., 32
De Broglie, and theory of light, 54, 55
Descartes, René, 15, 16
Documentation, 66

Einstein, 30
Entropy, 30
Euclid, 9
Eudoxus, 5
Extension, of a set, 35

Fermat, Pierre de, 16, 19
Fisher, R. A., 29, 30

Galois, E., 25
Gauss, Karl, 23
Gilbert, Geoffrey, 22
Gödel, K., 30, 31, 32
Group theory, 26

Hale, Matthew, 21, 22
Heuristics, 63–4, 67
Hippocrates of Chios, 5

Implication (partial and total), 39
Inclusion (partial and total), 39
Induction, 49, 50
Intension of a set, 35, 41

Judicial proof, plausibility of, 71

Lateral thinking, 65
Least squares, theory of, 23
Legendre, A-M, 23
Leibniz, 12, 17, 18, 19, 21
Locke, John, 22
Lull, Ramon, 11, 12

Maurolico, Francesco, 13
Medical diagnosis, plausibility of, 70, 71
Menaechmus, 5
Mendeleev, D., 28, 29

Newton, 19, 20

Osborn, A. F., 32

Parmenides, 3
Pascal, Blaise, 17
Pearson, K., 30
Philip of Macedon, 5, 6
Plato, 4, 5, 6, 8
Plausibility, definition of, 36, 38
Plausible exploration, method of, 63
Polya, G., 57, 58, 59, 60, 71

Ramus, Peter, 13
Russell, Bertrand, 30

Socrates, 3, 4
Speusippus, 4, 5, 6
Syllogism, 7, 27, 44, 45, 46

Theophrastus, 8
Theudius, 5
Transitivity, 41

Wilkins, John, 21, 22

Yates, F., 29

Zwicky, F., 31, 32